Methods in Microbiology
Volume 39

Recent titles in the series

Volume 24 *Techniques for the Study of Mycorrhiza*
JR Norris, DJ Reed and AK Varma

Volume 25 *Immunology of Infection*
SHE Kaufmann and D Kabelitz

Volume 26 *Yeast Gene Analysis*
AJP Brown and MF Tuite

Volume 27 *Bacterial Pathogenesis*
P Williams, J Ketley and GPC Salmond

Volume 28 *Automation*
AG Craig and JD Hoheisel

Volume 29 *Genetic Methods for Diverse Prokaryotes*
MCM Smith and RE Sockett

Volume 30 *Marine Microbiology*
JH Paul

Volume 31 *Molecular Cellular Microbiology*
P Sansonetti and A Zychlinsky

Volume 32 *Immunology of Infection, 2nd edition*
SHE Kaufmann and D Kabelitz

Volume 33 *Functional Microbial Genomics*
B Wren and N Dorrell

Volume 34 *Microbial Imaging*
T Savidge and C Pothoulakis

Volume 35 *Extremophiles*
FA Rainey and A Oren

Volume 36 *Yeast Gene Analysis, 2nd edition*
I Stansfield and MJR Stark

Volume 37 *Immunology of Infection*
D Kabelitz and SHE Kaufmann

Volume 38 *Taxonomy of Prokaryotes*
Fred Rainey and Aharon Oren

Methods in Microbiology
Volume 39

Systems Biology of Bacteria

Edited by

Colin Harwood
*Institute of Cell and Molecular Biosciences
Baddiley-Clark Building
Newcastle University
Newcastle upon Tyne
NE2 4AX*

Anil Wipat
*School of Computing Science
Claremont Tower
Newcastle University
Newcastle upon Tyne
NE1 7RU*

AMSTERDAM • BOSTON • HEIDELBERG • LONDON
NEW YORK • OXFORD • PARIS • SAN DIEGO
SAN FRANCISCO • SINGAPORE • SYDNEY • TOKYO
Academic Press is an imprint of Elsevier

Academic Press is an imprint of Elsevier
The Boulevard, Langford Lane, Kidlington, Oxford, OX51GB, UK
32, Jamestown Road, London NW1 7BY, UK
Radarweg 29, PO Box 211, 1000 AE Amsterdam, The Netherlands
225 Wyman Street, Waltham, MA 02451, USA
525 B Street, Suite 1900, San Diego, CA 92101-4495, USA

First edition 2012

Copyright © 2012 Elsevier Ltd. All rights reserved

No part of this publication may be reproduced, stored in a retrieval system or transmitted in any form or by any means electronic, mechanical, photocopying, recording or otherwise without the prior written permission of the publisher.

Permissions may be sought directly from Elsevier's Science & Technology Rights Department in Oxford, UK: phone (+44) (0) 1865 843830; fax (+44) (0) 1865 853333; email: permissions@elsevier.com. Alternatively you can submit your request online by visiting the Elsevier web site at http://elsevier.com/locate/permissions, and selecting Obtaining permission to use Elsevier material.

Notice

No responsibility is assumed by the publisher for any injury and/or damage to persons or property as a matter of products liability, negligence or otherwise, or from any use or operation of any methods, products, instructions or ideas contained in the material herein. Because of rapid advances in the medical sciences, in particular, independent verification of diagnoses and drug dosages should be made.

ISBN: 978-0-08-099387-4
ISSN: 0580-9517 (Series)

For information on all Academic Press publications
visit our website at www.store.elsevier.com

Printed and bound in UK

12 13 14   10 9 8 7 6 5 4 3 2 1

Contents

Series Advisors xi
Contributors xiii

**CHAPTER 1 High-Resolution Temporal Analysis of Global
Promoter Activity in *Bacillus subtilis* 1**
Eric Botella, David Noone, Letal I. Salzberg,
Karsten Hokamp, Susanne Krogh Devine, Mark Fogg,
Anthony J. Wilkinson, Kevin M. Devine

1. Introduction 1
2. Gene Fusion Technology 2
   2.1 History 2
3. Reporter Proteins 3
   3.1 β-Galactosidase 3
   3.2 Green Fluorescent Protein 3
   3.3 Luciferase 4
4. Cellular Placement of Transcriptional Fusions 5
5. Choice of System 6
6. Use of Promoter Fusions to Study Biological Processes
   at High Temporal Resolution in Bacteria 6
   6.1 Analysis of Promoter Activity in *E. coli* and
       Salmonella Species 6
   6.2 Analysis of Promoter Activity in *B. subtilis* 9
7. Methodology for High-Throughput Analysis of Promoter
   Activity with Fine Temporal Resolution in *B. subtilis* 12
   7.1 Promoter Fragment Selection and Generation 12
   7.2 Plasmids 13
   7.3 High-Throughput Cloning and Plasmid Preparation 15
   7.4 Strain Storage 16
   7.5 Growth and Data Collection 19
   7.6 Data Analysis 20
   7.7 Visualization of Promoter Activity 21
8. Conclusions 22
   Acknowledgements 22
   References 24

CHAPTER 2 Data Mining for Microbiologists 27
J.S. Hallinan

1. Introduction 27
2. What is Data Mining? 28
3. The Data Mining Process 30
4. Inferential Techniques 32
   - 4.1 Basic Statistical Analysis 32
   - 4.2 Linear Regression 33
   - 4.3 Discriminant Analysis 33
   - 4.4 Principal Components Analysis 35
5. Machine Learning Techniques 37
   - 5.1 Support Vector Machines 37
   - 5.2 Hidden Markov Models 38
   - 5.3 Decision Trees 40
   - 5.4 Clustering 42
   - 5.5 Artificial Neural Networks 48
   - 5.6 Ontologies and Text Mining 50
   - 5.7 Data Integration and Network Analysis 56
6. The Role of eScience 59
7. Case Study: Data Mining for Protein Function Prediction 63
8. Data Mining with Microbial Data: Practical Issues 65
   - 8.1 Noise 65
   - 8.2 Overfitting 66
   - 8.3 The Peaking Phenomenon 67
9. Conclusions 68
   - References 68

CHAPTER 3 Proteomics: From Relative to Absolute Quantification for Systems Biology Approaches 81
Andreas Otto, Jörg Bernhardt, Michael Hecker, Uwe Völker, Dörte Becher

1. Introduction 81
   - 1.1 'Omics' Techniques in Systems Biology Approaches 81
   - 1.2 Gel-Based Proteomics 81
   - 1.3 Gel-Free Proteomics 82
   - 1.4 Targeted Proteomics 83
   - 1.5 Proteomics Based on Data-Independent Acquisition 83
2. Absolute Quantification Workflows in Proteomics 84
   - 2.1 Fusion Protein-Based Global Scale Protein Quantification 84

2.2 Mass Spectrometry-Based Absolute Quantification
Workflows in Proteomics 85
2.3 Large-Scale Absolute Quantification in Proteomics 87
3. The Proteomics Workflow 87
3.1 Sample Preparation for Absolute Quantification Strategies
in Proteomics 87
4. Generation of Absolute Quantitative Data by Targeted Mass
Spectrometry 92
4.1 Criteria for Selection of Peptides for Targeted Analyses 92
4.2 Basic Requirements for AQUA 94
4.3 Quantitative SRM Assays Based on Heavy-Labelled
Peptides 96
5. Generation of Large-Scale Relative Proteomics Data:
Differential 2D Gel Image Analysis 97
6. Large-Scale Absolute Quantitative Proteomics with
SRM-Calibrated 2D PAGE 99
References 101

CHAPTER 4 Imaging Fluorescent Protein Fusions in Live Bacteria 107
Geoff Doherty, Karla Mettrick, Ian Grainge, Peter J. Lewis

1. Introduction 107
2. Molecular Toolkits 107
   2.1 *Bacillus subtilis* 108
   2.2 *Escherichia coli* 113
   2.3 *Staphylococcus aureus* 116
   2.4 *Acinetobacter* spp. 116
   2.5 Tet Array and pLau53 117
3. Functional Analysis 117
4. Growth Conditions 117
   4.1 Liquid 118
   4.2 Solid 118
5. Agarose Slide Preparation 119
6. Imaging Hardware 119
   6.1 Microscope and Camera 119
   6.2 Objectives 120
   6.3 Filters 120
7. Imaging Software 121
   7.1 Image Processing Software 121
8. Image Processing 122
   8.1 Signal Comparison 123

9. Concluding Remarks 124
   Acknowledgements 124
   References 124

CHAPTER 5 Targeted and Quantitative Metabolomics in Bacteria 127
Hannes Link, Joerg Martin Buescher, Uwe Sauer

1. Introduction 127
2. Cultivation of Bacteria for Metabolome Analysis 128
   2.1 Medium 129
   2.2 Inoculum 130
   2.3 Cultivation 130
3. Sampling Bacterial Cultures for Metabolome Analysis 131
   3.1 Fast Filtration 132
   3.2 Whole Cell Broth Extraction 133
   3.3 High-Throughput Sampling of Bacterial Cultures in 96-Well Format 134
4. Extraction of Metabolites from Bacteria 135
   4.1 Hot Ethanol Extraction of Filtered Samples 135
   4.2 Hot Ethanol Extraction of Whole Cell Broth 136
   4.3 Acidic Acetonitrile Extraction of Filtered Samples 136
5. Analysis by Mass Spectrometry 137
   5.1 Separation by Ion-Pairing Chromatography 138
   5.2 Detection by ESI-MS/MS 139
   5.3 Peak Integration of LC–MS/MS Data 140
   5.4 Isotope Ratio-Based Quantification of Intracellular Metabolite Concentrations 141
   5.5 Normalizing Quantified Metabolites to Biomass or Total Cell Volume 145
6. Interpretation of Metabolite Data 146
7. Outlook 147
   References 148

CHAPTER 6 Array-Based Approaches to Bacterial Transcriptome Analysis 151
Ulrike Mäder, Pierre Nicolas

1. Introduction 151
2. Prior Considerations 153
   2.1 Technical Requirements for Microarray Experiments 153
   2.2 Experimental Design 156
3. Performing the Experiments 158

    3.1 RNA Preparation and Quality Assessment 158
    3.2 Synthesis and Labelling of cDNA 160
    3.3 Array Hybridization, Scanning and Data Extraction 164
4. Data Analysis 165
    4.1 Data Preprocessing and Quality Checks 165
    4.2 Expression and Differential Expression 169
    4.3 Towards Gene Expression Networks 173
5. Final Comments 177
    References 177

Index **183**

Series Advisors

Gordon Dougan
The Wellcome Trust Sanger Institute, Wellcome Trust Genome Campus, Hinxton, Cambridge CBIO ISA, UK

Graham J Boulnois
Schroder Ventures Life Science Advisers (UK) Limited, 71 Kingsway, London WC2B 6ST, UK

Jim Prosser
School of Medical Sciences, University of Aberdeen, Cruickshank Building, St Machar Drive, Aberdeen, AB24 3UU, UK

Ian R Booth
School of Medical Sciences, University of Aberdeen, Institute of Medical Sciences, Foresterhill, Aberdeen AB25 2ZD, UK

David A Hodgson
Department of Biological Sciences, University of Warwick, Conventry CV4 7AL, UK

David H Boxer
University of Dundee, Dundee DD1 4HN, UK

Contributors

Dörte Becher
Ernst-Moritz-Arndt-University Greifswald, Institute for Microbiology, Greifswald, Germany

Jörg Bernhardt
Ernst-Moritz-Arndt-University Greifswald, Institute for Microbiology, Greifswald, Germany

Eric Botella
Smurfit Institute of Genetics, Trinity College Dublin, Dublin, Ireland

Joerg Martin Buescher
Institute of Molecular Systems Biology, ETH Zurich, Zurich, Switzerland, and Biotechnology Research and Information Network AG, Zwingenberg, Germany

Kevin M. Devine
Smurfit Institute of Genetics, Trinity College Dublin, Dublin, Ireland

Susanne Krogh Devine
Smurfit Institute of Genetics, Trinity College Dublin, Dublin, Ireland

Geoff Doherty
School of Environmental and Life Sciences, University of Newcastle, Callaghan, New South Wales, Australia

Mark Fogg
Structural Biology Laboratory, Department of Chemistry, University of York, York, United Kingdom

Ian Grainge
School of Environmental and Life Sciences, University of Newcastle, Callaghan, New South Wales, Australia

J.S. Hallinan
School of Computing Science and Centre for Bacterial Cell Biology, Newcastle University, Newcastle upon Tyne, United Kingdom

Michael Hecker
Ernst-Moritz-Arndt-University Greifswald, Institute for Microbiology, Greifswald, Germany

Karsten Hokamp
Smurfit Institute of Genetics, Trinity College Dublin, Dublin, Ireland

Peter J. Lewis
School of Environmental and Life Sciences, University of Newcastle, Callaghan, New South Wales, Australia

Hannes Link
Institute of Molecular Systems Biology, ETH Zurich, Zurich, Switzerland

Ulrike Mäder
Ernst-Moritz-Arndt-University Greifswald, Interfaculty Institute for Genetics and Functional Genomics, Greifswald, Germany

Karla Mettrick
School of Environmental and Life Sciences, University of Newcastle, Callaghan, New South Wales, Australia

Pierre Nicolas
INRA, UR1077 Mathématique, Informatique et Génome (MIG), Domaine de Vilvert, Jouy-en-Josas, France

David Noone
Smurfit Institute of Genetics, Trinity College Dublin, Dublin, Ireland

Andreas Otto
Ernst-Moritz-Arndt-University Greifswald, Institute for Microbiology, Greifswald, Germany

Letal I. Salzberg
Smurfit Institute of Genetics, Trinity College Dublin, Dublin, Ireland

Uwe Sauer
Institute of Molecular Systems Biology, ETH Zurich, Zurich, Switzerland

Uwe Völker
Ernst-Moritz-Arndt-University Greifswald, Interfaculty Institute for Genetics and Functional Genomics, Greifswald, Germany

Anthony J. Wilkinson
Structural Biology Laboratory, Department of Chemistry, University of York, York, United Kingdom

CHAPTER 1

High-resolution temporal analysis of global promoter activity in *Bacillus subtilis*

Eric Botella*,[1], David Noone*, Letal I. Salzberg*, Karsten Hokamp*, Susanne Krogh Devine*, Mark Fogg[†], Anthony J. Wilkinson[†], Kevin M. Devine*,[1]

*Smurfit Institute of Genetics, Trinity College Dublin, Dublin, Ireland
[†]Structural Biology Laboratory, Department of Chemistry, University of York, York, United Kingdom
[1]Corresponding authors. e-mail address: botellae@tcd.ie; kdevine@tcd.ie

1 INTRODUCTION

Until recently, the reductionist approach predominated in biological research. Its objectives were to describe and understand a biological entity in terms of its subsystems, their constituent components and the molecular interactions both within and between them. This approach emerged from the prevailing technological capabilities, whereby single or small numbers of molecules and the interactions between them could be described in great detail. However, this approach is limited in that even the simplest biological system is highly complex, composed of a myriad of subsystems and their constituent components that interact in a highly precise and regulated way. A key concept of such systems is emergence: the idea that integration of the subsystems leads to the emergence of novel properties and characteristics that cannot be predicted from the analysis of the individual constituent parts. Just as the functioning and uses of a car cannot be adequately described or understood in terms of its constituent parts and subsystems, so it is that understanding a biological system requires a more global and holistic approach to reveal novel characteristics. To some extent, "life" is the emergent property of the sum of the components and specific interactions that occur in the cell or organism. It is evident that global approaches are required to better describe biological systems and how they change in response to alterations in the prevailing conditions.

A key objective, therefore, is to establish the regulatory circuitry and dynamics of gene expression at a global level. Many attributes of the regulatory circuitry can be deduced from the changes that occur in cellular RNA profiles in response to diverse stimuli. Techniques such as RNA-Seq and high-density microarrays give a very detailed static view of the spectrum of cellular RNA species and their prevalence at any particular time point, from which many features of the regulatory network can be inferred (Chechik et al., 2008; Chechik and Koller, 2009; Ralser et al., 2009). However, the complexity and cost of RNA-Seq and microarrays makes them unsuited to establishing a dynamic model of differential gene expression at high temporal resolution. Promoter fusion technology, however, allows dynamic changes in gene expression to be established at high temporal resolution with relative ease and at low cost. This approach involves establishing promoter activity by monitoring expression of the *gfp* and *lux* reporter genes. Green fluorescent protein (GFP) from *Aequorea victoria* and the bioluminescent firefly and bacterial luciferase (Lux) systems have revolutionized such expression studies. Cellular levels of GFP and Lux can be established without perturbing the system under study. The activity of promoters can be established at high temporal resolution because of the cellular half-lives of GFP (very long) and Lux (very short). Moreover, promoter fusions can be generated in an automated high-throughput manner, allowing gene expression profiles to be established on a global scale.

In this chapter, we summarize the essential features of the GFP and Lux reporter proteins. We outline how they have been applied to establishing global changes in promoter activity at high temporal resolution in *Escherichia coli* and *Bacillus subtilis*. We then describe a methodology we have developed to generate promoter fusions in a high-throughput manner and establish gene expression on a global scale with high reproducibility in *B. subtilis* that can easily be adapted for expression studies in other bacteria.

2 GENE FUSION TECHNOLOGY
2.1 History

The generation and use of reporter gene fusions has a long history in molecular genetics (for review, see Silhavy and Beckwith, 1985). Some of the first fusions were generated by Benzer in his studies on the independently transcribed *r*IIA and B genes of the T4 phage. A deletion extending from the carboxy end of the A gene to the amino end of the B gene removed all expression signals of the B gene and generated a translational fusion where the hybrid *r*IIA-B protein retained some B activity (Champe and Benzer, 1962). Mutations of the A gene could now be studied by measuring their effect on B protein activity (Silhavy and Beckwith, 1985). This established the fundamental principle of gene fusion technology, whereby the expression signals of any gene can be studied by fusing them to a reporter gene whose product is easily assayable. Fusion technology was further developed by combining knowledge of the *lac* operon with that of mobile genetic elements such as phages and transposons. LacZ emerged as a reporter of choice because β-galactosidase activity could easily be assayed using substrates that yielded coloured compounds (Slauch and Silhavy, 1991). Thus a promoterless *lacZ* reporter gene was combined with modified Mu and λ phages [e.g. Mu d1(*Ap lac*) and λ *plac* Mu] that allowed the random generation of transcriptional fusions with chromosomal promoters (for review, see Silhavy and Beckwith, 1985). The advent of recombinant DNA technology brought precision and sophistication to the construction of fusion generating vectors. Plasmids and transposons replaced the more cumbersome phages as the genetic element of choice for the construction and mobilization of fusions. Transcriptional and translational fusions could be rapidly generated with precision using the polymerase chain reaction (PCR). Fusion technology could be extended to a wide range of bacteria by adapting existing plasmids for use in new hosts. This was achieved by altering the expression signals of antibiotic-resistance genes (required for selection) and the ribosome-binding sites of reporter genes for optimal function in the new host. Fusions could be established on replicating plasmids or integrated into the chromosome at homologous or neutral heterologous (ectopic) sites. The disadvantage of having to prepare cell lysates in order to quantitate LacZ levels was overcome by using GFP (encoded by the *gfp* gene) and luciferase (encoded by the *lux* genes) whose intracellular levels can be measured in real time without perturbing the cells under study (see Section 3). Furthermore, constructing *gfp* and *lux* promoter fusions is particularly amenable to high-throughput technology. Generating PCR-generated promoter fragments of any size, inserting them into *gfp*- or *lux*-containing vectors by ligation-independent cloning (LIC) and transforming them into *E. coli* or other host strains can be easily automated. Strains containing promoter fusions can be grown in a 96-well plate format with optical density, fluorescence or luminescence monitored automatically in time intervals as short as 1 min. The feasibility of using fusion technology to establish promoter activity on a global scale is illustrated by the construction of a comprehensive library of $\sim$2000 promoter GFP fusions in *E. coli* K12 and monitoring their expression during a diauxic glucose–lactose shift (Zaslaver et al., 2006). Moreover, once constructed, the library of promoter fusions can be used in whole or in part to investigate gene expression under a myriad of environmental, nutritional or developmental conditions (Zhang et al., 2011). Such technical developments have created a robust and flexible technology capable of rapidly and reproducibly establishing dynamic changes in promoter activity with high temporal resolution at a relatively low cost.

3 REPORTER PROTEINS
3.1 β-Galactosidase

The *lacZ* gene encoding β-galactosidase from *E. coli* was the most widely used reporter in both prokaryotic and eukaryotic model systems. While proving to be a robust and reliable reporter protein, LacZ had several limitations. Many bacteria have endogenous β-galactosidase activity. The LacZ protomer is very large ($\sim$120 kDa) requiring more than 3 kb of DNA to encode it so that plasmids encoding *lacZ* fusions are correspondingly large and often prone to deletion. While LacZ is a good qualitative indicator of expression, quantitation requires cell harvesting and lysate preparation, features requiring the biological system under study to be sampled and perturbed. This makes LacZ unsuitable for high-throughput analysis and high definition monitoring of gene expression.

3.2 Green fluorescent protein

Use of LacZ has been largely superseded by the GFP from the jellyfish *A. victoria* (for review, see Tsien, 1998). The singular property that makes GFP the reporter of choice for establishing promoter activity is the ability to self-generate the chromophore during folding and maturation without the requirement for any additional molecule and to do so in a range of organisms. The chromophore is generated by maturation of the side chains of amino acids 65–67 that are located within an 11-stranded β-barrel structure. By changing amino acids at positions 65–67, GFP variants with increased brightness and shifted excitation and emission spectra have been generated (Cormack et al., 1996; Scholz et al., 2000). Some of the resulting palette of coloured variants, such as cyan (CFP) and yellow (YFP) fluorescent proteins, can be distinguished spectrophotometrically, allowing simultaneous visualization of several promoter activities in the same cell (Stepanenko et al., 2011). The folding kinetics and stability of GFP and its variants are features crucial to their use in determining real-time promoter activity. Several fast-folding variants now exist in which the chromophore is rapidly generated after GFP expression. This ensures that measurement of fluorescence at high temporal resolution accurately reflects promoter activity (Cormack et al., 1996). High temporal resolution of promoter activity also requires the ability to precisely measure the amount of reporter protein produced in short-time intervals. This is achieved most accurately when the biological half-life of the reporter protein is either significantly shorter (e.g. Lux) or longer (e.g. GFP) than the duration of the time interval within which expression is measured. GFPmut3 is a fast-folding variant that is extraordinarily stable in *E. coli* and *B. subtilis* with estimated half-lives of more than 24 h and approximately 10 h, respectively (Andersen et al., 1998; Botella et al., 2011). Therefore, it is particularly suited to accurate measurement of promoter activity in bacterial systems where growth cycles and promoter response times are considerably shorter than these half-life values.

Several additional features of GFP are worthy of consideration when choosing it as a reporter for high-resolution kinetic analysis of promoter activity. (1) The excitation wavelength required to generate GFP fluorescence often also generates significant autofluorescence in biological material, resulting in a high background. This can result in problems of low signal-to-noise ratios, especially when analyzing less active promoters. (2) Many molecules of biological interest (e.g. antibiotics) fluoresce at the wavelengths used to excite GFP, a feature that will complicate or even prevent its use as a reporter protein when studying their effects on gene expression. (3) Formation of the GFP chromophore requires molecular oxygen, which makes it of limited use in obligate anaerobic organisms. (4) GFP excitation can generate significant quantities of intracellular hydrogen peroxide, and cells expressing this protein at high concentrations may be oxidatively stressed (Tsien, 1998).

3.3 Luciferase

Luciferase is a second reporter used for high-resolution global analysis of promoter activity. Two types of luciferase protein are commonly used, firefly (*Photinus pyralis*) and bacterial. Firefly luciferase uses luciferin as a substrate, oxidizing it to oxyluciferin in a reaction that utilizes molecular oxygen and ATP, and liberates light at 560 nm (Wilson and Hastings, 1998; Fraga, 2008). Bacterial luciferase is encoded by the *luxCDABE* operon that is usually sourced from *Photorhabdus luminescens*, *Vibrio harveyi* or *Vibrio fischeri*. Bacterial luciferase catalyzes the oxidation of reduced flavin mononucleotide (FMNH$_2$) and myristyl aldehyde to myristic acid and FMN, a reaction that liberates light at 490 nm (Waidmann *et al.*, 2011). The *luxAB* genes encode the heterodimeric luciferase while the *luxCDE* genes encode the enzymes required for generation of the myristyl aldehyde substrate from myristol ACP (Waidmann *et al.*, 2011). An important consideration when choosing the type of luciferase to use is that the firefly enzyme requires exogenous addition of the decanal substrate, whereas the bacterial system can be engineered (by inclusion of *luxCDE* in the operon) to generate the substrate endogenously. The suitability of luciferase systems for determining promoter activity in real time resides in the instability of the enzyme. Firefly luciferase is unstable in *B. subtilis* with a half-life of only ∼6 min (Mirouze *et al.*, 2011a,b). Thus the suitability of GFPmut3 (half-life of ∼10 h) and firefly luciferase (half-life or ∼6 min) for high-resolution kinetic analysis of promoter activity emanates from their contrasting stabilities *in vivo*. A particular advantage of luciferases is that most cells are not luminescent so that high signal-to-noise ratios can be achieved. This makes them particularly suited to the analysis of promoters with low activity. The necessity for molecular oxygen to generate luminescence is a disadvantage of luciferase reporters, essentially precluding their use as reporters in obligate anaerobic bacteria. The usage of energy (ATP) and reducing equivalents (FMNH$_2$) by firefly and bacterial luciferases, respectively, might also complicate the measurement or interpretation of promoter activity in metabolic studies.

Despite these limitations, both GFP and luciferases are versatile reporter proteins that are widely used in expression studies to determine promoter activity. Both allow global promoter activity to be established with high temporal resolution and reproducibility without the need to sample or perturb the system under study in any way.

4 CELLULAR PLACEMENT OF TRANSCRIPTIONAL FUSIONS

Placement of the fusion within the cell is an important consideration when planning a global study of promoter activity. Fusions can be placed on replicating plasmids or integrated into the chromosome at homologous or heterologous locations. Placing the fusion on a plasmid makes it easy to mobilize between strains but has the potential disadvantages of plasmid loss through structural or segregational instability and of variation in copy number, all of which will influence the level of promoter activity. Alternatively, fusions can be integrated into the chromosome by a Campbell-type single crossover event at the locus homologous to the promoter region under study. A DNA fragment containing the entire promoter region with its 3′-end located upstream of the ribosome binding site of the first gene of the operon under study is usually chosen as the promoter-containing fragment (Botella *et al.*, 2010, 2011). This type of integration is particularly suited to high-throughput generation and analysis of promoter fusions where plasmid construction and transformation can be automated. A limitation of this approach is that Campbell-type integrations can result in the promoter fusion being present in multiple copies on the chromosome. This can be caused by amplification of the integrated plasmid through unequal crossing-over at the duplicated sequences during chromosomal replication or by integration of a multimeric plasmid (Janniere et al., 1985; Young, 1984). This results in individual transformants having similar expression profiles but different levels of promoter activity (usually multiples of 2–4×, E Botella and KM Devine, unpublished). The most desirable placement of a promoter fusion is at a neutral heterologous locus by a double-crossover event. A locus such as *amyE* of *B. subtilis* (that encodes α-amylase) is useful in that successful integration can be monitored by loss of α-amylase production with a simple iodine–starch plate assay. Promoter fusions placed on the chromosome by this mechanism are stable and present in single copy. Chromosomal integration of promoter fusions by a double-crossover event is best achieved with linear DNA fragments, generated either by linearizing plasmids or by overlapping PCR. This approach is less amenable to high-throughput analysis because of the additional DNA manipulations required and because double-crossover integration events occur at significantly lower frequencies than do single-crossover events.

5 CHOICE OF SYSTEM

Undertaking a global study of promoter activity in an organism requires careful consideration of the system most appropriate to fulfill the objectives of the study. Accurate determination of promoter activity requires that the reporter protein folds rapidly and be either highly stable or unstable in the organism under study. While GFPmut3 (stable) and firefly luciferase (unstable) have these properties in *B. subtilis*, they may behave differently in other bacterial species. When choosing between GFP and luciferases, it is important to evaluate the likely strengths of the promoters under study: very high-level production of GFP may be toxic, while very low-level production may not be detectable over background autofluorescence. Luciferases are more suited to studying low-activity promoters because of the higher signal-to-noise ratios. However, firefly luciferase requires the addition of substrate, while bacterial luciferases require additional genes to generate its substrate intracellularly, which makes plasmid constructs more complex. The decision on where to place the promoter fusion is a critical one with an inverse relationship between the ease of construction and avoidance of potential artefacts. Integration of promoter fusions into a neutral chromosomal locus is optimal but more difficult to achieve than integration by a single crossover or establishing the promoter fusions on a replicating plasmid. However, the latter approach can give artefactual results in some instances.

6 USE OF PROMOTER FUSIONS TO STUDY BIOLOGICAL PROCESSES AT HIGH TEMPORAL RESOLUTION IN BACTERIA

6.1 Analysis of promoter activity in *E. coli* and *Salmonella* species

Promoter activity analyses in *E. coli* and *Salmonella typhimurium* have tended to use *gfpmut2*, *gfpmut3* and *lux* reporter genes located on low copy number (pSC101 derivatives) replicating plasmids. Two approaches can be used to generate promoter fusions. Goh *et al.* (2002) employed a random approach by inserting chromosomal DNA fragments into a low copy number plasmid encoding a promoterless *lux* reporter gene and screening transformants for altered promoter activity upon exposure to subinhibitory antibiotic concentrations. A large number of promoters were identified with a significant number showing altered activity upon exposure of cells to subinhibitory antibiotic concentrations. Use of the *lux* reporter system in this study eliminated the possibility that the antibiotics themselves would interfere with the analysis of expression by fluorescing in response to the excitation light source.

Bjarnason *et al.* (2003) used the *lux* genes from *P. luminescens* to identify iron responsive promoters in *Salmonella enterica*. They generated a random promoter library by cloning *Sau*3A-generated chromosomal DNA fragments in plasmid pCS26-Pac and identified iron-responsive promoters by comparing expression of the strain library growing on media with low and high iron content. Three classes of iron-responsive genes were identified, two of which were regulated by Fur. The third class was Fur independent and contained promoters that responded both negatively and positively to iron availability.

Alon and colleagues have employed a directed approach, cloning specific promoter-containing fragments into low copy number plasmids (pSC101 replicon) encoding promoterless *gfpmut2* or *gfpmut3* reporter genes. Their studies focused on the constituent promoters for regulatory networks such as flagella synthesis (Kalir *et al.*, 2001), the SOS regulon (Ronen *et al.*, 2002) and amino acid biosynthetic operons (Zaslaver *et al.*, 2004), culminating in the construction of a comprehensive library of ~2000 transcriptional fusions for *E. coli* (Zaslaver *et al.*, 2006).

There are three classes of genes involved in flagellar synthesis: Class I genes encode the regulator FlhCD; Class II genes encode the basal body and hook of the flagella and the FliA and FlgM regulators that control expression of the class III genes. Temporal regulation is conferred by interaction between the FlgM and FliA proteins that inactivates FliA regulatory activity. However, FlgM is exported from the cell upon formation of the basal body and hook structure, thereby liberating FliA to activate the Class III genes. High-resolution expression profiles of flagellar proteins, established using *gfpmut3* transcriptional fusions, were consistent with previous reports (Kalir *et al.*, 2001). The temporal synchronization of gene expression with the assembly of flagella and the onset of chemotaxis is clearly shown: basal body and hook proteins are expressed first, followed by flagellar proteins and then by the chemotaxis system. The authors point out that this temporal order is probably not strictly necessary, as functional flagella are formed in cells where some constituent proteins are expressed from heterologous promoters. However, this temporal sequence may contribute to the efficiency with which flagella are produced in growing cells (Kalir *et al.*, 2001).

A separate study established the expression kinetics of the constituent genes of the SOS DNA repair system of *E. coli* at high temporal resolution (Ronen *et al.*, 2002). The SOS response is composed of approximately 30 operons whose expression is negatively regulated by the LexA repressor. DNA damage, detected by the generation of single-stranded DNA, activates RecA leading to cleavage of the LexA repressor and derepression of the constituent genes of the LexA regulon. Ronen et al. (2002) fused the promoter regions of eight LexA regulon genes (*uvrA*, *uvrD*, *lexA*, *recA*, *ruvA*, *polB*, *umuD* and *uvrY*) to a promoterless *gfpmut3* gene and established the fusion constructs in *E. coli* on a low copy number (pSC101-based replicon) plasmid. Several important features of the dynamics of SOS response induction emerged from this study. The activities of all eight promoters increased within minutes of UV light exposure, each with a distinctive kinetic profile and expression level. The kinetic profiles were consistent with a less-temporally refined transcriptomic analysis, supporting the accuracy of this reporter fusion-based study (Courcelle et al., 2001). Promoter activities increased for approximately 20 min and then decreased with distinctive kinetics. The order in which the genes were turned off reflected their function in SOS-mediated DNA repair: the expression of genes encoding nucleotide excision repair functions (*uvrA*) and the RecA and LexA regulators (*recA lexA*) was turned off first, while genes (*polB*, *uvrD* and *ruvA*) encoding proteins involved in other repair processes of the SOS response were turned off later. Thus a temporal correlation was shown to exist between the kinetics of promoter activity and the order of events during DNA repair and the restoration of DNA replication. The authors also showed a close correlation between the level of active LexA protein (determined experimentally by Western blotting analysis) and that inferred from the activities of the constituent regulon promoters. Thus promoter activity is a good indicator of the activity of transcriptional regulators *in vivo*.

A subsequent study from the Alon group established promoter activities for operons that encode the enzymes for several amino acid biosynthetic pathways. The rationale for this study is that each amino acid biosynthetic pathway constitutes an individual regulatory network. Thus by determining promoter activities for operons that encode the enzymes of each amino acid biosynthetic pathway at high temporal resolution, it is possible to establish regulatory principles for how these biosynthetic pathways are controlled and how they are networked. The activities of 52 promoter fusions were established at high temporal resolution in this study. Promoter-containing fragments were cloned into a low copy number (pSC101-based replicon) plasmid containing promoterless *gfpmut2* (pUA66) or *luxCDABE* (pUAL94) genes. Several novel features of the regulation of amino acid biosynthesis were revealed. As expected, promoters are generally turned off when all amino acids are present in the medium and turned on when specific amino acids are absent. Addition of one amino acid generally leads to repression of the promoters of the operons that encode its synthesis. However, cross-regulation was also observed: in some cases, addition of a particular amino acid leads to altered expression of operons that encode the synthetic enzymes of other amino acids. Thus promoter fusion analysis reveals the existence of a complex web of cross-regulation of amino acid biosynthesis in *E. coli*. Such analyses also reveal the temporal sequence of promoter activation that occurs when a particular amino acid is removed from the medium. Promoters are activated in a sequence that sees those for genes encoding enzymes acting early in the pathway being turned on first, while those for genes encoding enzymes acting later in the pathway are turned on up to 10 min later (Zaslaver et al., 2004).

These studies were extended by construction of a comprehensive library of transcriptional promoter fusions (~2000) for *E. coli* (Zaslaver et al., 2006). Intergenic regions were amplified by PCR and cloned into plasmid pUA66 (low copy number pSC101-based derivative) that encodes a promoterless *gfpmut2*. GFPmut2 is a fast-folding variant of GFP that becomes fluorescent within 5 min of expression and is highly stable in *E. coli*. The authors established the kinetics of promoter activity for this library of transcriptional fusions during a diauxic shift, monitoring GFP production at 7-min intervals over a period of 20 h. Cells were grown in medium containing limiting glucose (0.03%) and non-limiting lactose, conditions under which *E. coli* utilizes glucose first and then lactose. More than 60% of the promoter fusions showed activity under these conditions with many genes being strongly expressed during the early and mid-logarithmic growth phases. Expression of the *lac* and *gal* operons was particularly interesting, as the activity of their promoters increased before growth of the culture slows down and peaked at the point of growth rate reduction.

This library of transcriptional fusions has been utilized by the Alon group to investigate several additional features of promoter activity in *E. coli*. They have investigated the distribution of promoter activities in cells growing at different growth rates and under different growth conditions (Zaslaver et al., 2009). A particularly interesting study examined the variance of promoter activity in cell populations by flow cytometry for the ~2000 fusion-containing strains in their library (Silander et al., 2012). It has been widely observed that some genes show significant fluctuations in expression in individual cells within a population. Such differences in expression level (called phenotypic noise) can result in different phenotypes being manifest within a cell population. Important observations are that the promoter alone can contribute to the generation of phenotypic noise while essential and highly conserved genes have lower levels of phenotypic noise than do non-essential and lowly conserved genes (Silander et al., 2012).

The library of transcriptional promoter fusions generated by the Alon group was also used to establish the toxicity of naphthenic acids (cyclopentyl and cyclohexyl carboxylic acids) that are components of effluents from oil production facilities (Zhang et al., 2011). The library of strains with promoter *gfpmut2* fusions was exposed to the range of naphthenic acid concentrations typically observed in the environment, and changes in promoter activity profiles were established. Results showed that the main cellular response of *E. coli* to naphthenic acids was to increase expression of genes involved in the pentose phosphate pathway and to down-regulate expression of an ABC-type transporter complex. Exposure to naphthenic acids also affected adenylate cyclase metabolism and the SOS response.

6.2 Analysis of promoter activity in *B. subtilis*

Global analyses of promoter activities at high temporal resolution have also been performed in *B. subtilis* (Botella et al., 2011; Buescher et al., 2012). Transcriptional reporter fusions were generated in a high-throughput manner: promoter-containing DNA fragments were directionally inserted upstream of a promoterless *gfpmut3* gene, encoded on the integrating plasmid pBaSysBioII, by LIC cloning (Botella et al., 2010). This allowed transcriptional fusions to be established on the chromosome at the homologous locus by a non-mutagenic Campbell-type insertion (Botella et al., 2010). The GFPmut3 protein has a half-life of $\sim$10 h in B. subtilis, making it an ideal reporter for measuring promoter activities at high temporal resolution. The Botella et al. (2011) study focused on cell wall metabolism in wild-type and $\Delta phoPR$ mutant cells growing under three experimental conditions: a defined high phosphate medium (HPDM), under phosphate-limiting conditions (LPDM) and during phosphate replenishment of phosphate-limited cells in LPDM. GFP levels were measured at 10-min intervals and all experiments were performed with three independent clones of each promoter fusion, with the expression profile of each clone determined at least in duplicate using two different Biotek2 plate readers. This data set comprised a minimum of 150,000 data points. Expression profiles were visualized in heat map format, with promoter activity values normalized to the maximum observed under those conditions and expressed on a scale of 0–1. By documenting the numerical maximum activity attained by each promoter, the strengths of different promoters can be compared under these experimental conditions. A second important feature of this study is that promoter activities were compared with the RNA abundance of each operon determined using high-density microarrays. Discrepancies between promoter activities and RNA abundances (e.g. a high promoter activity with a low RNA abundance) can indicate post-transcriptional regulation. The very high degree of reproducibility with which promoter activity profiles and RNA levels were established is a notable feature of this study. Significant promoter activity ($\geq$25 units above the background threshold) was detected in 109 of the 135 cell wall-associated promoters examined. Several novel features of cell wall metabolism emerged from these studies that could only have been revealed by determining promoter activity at high temporal resolution. Of the 109 promoter activity profiles established, those of WalRK (YycFG)-controlled genes are unique in terms of their both kinetics and activity levels: they are the first genes to be turned on when stationary phase cells are diluted into fresh medium and are expressed at a significantly higher level than all other genes involved in cell wall metabolism. This study also revealed the temporal heterogeneity with which PhoPR regulon promoters are activated upon phosphate limitation (Botella et al., 2011). Of particular note is the early activation of the *tuaABCDEFGH* operon (encodes the enzymes for teichuronic acid biosynthesis) promoter that is turned on before the culture ceases exponential growth. Its promoter reaches maximum activity up to 30 min before the majority of the other PhoPR-controlled promoters (that control expression of genes encoding phosphate scavenging activities) are activated. This promoter profile suggests that teichuronic acid synthesis plays a distinctive role in the cellular response to phosphate limitation. This result also resonates with the diauxic shift study of Alon and colleagues who observed increased *lac* and *gal* promoter activities before growth slows as glucose level falls beneath the utilizable threshold (Zaslaver et al., 2006). A further notable feature of the *B. subtilis* response to phosphate limitation is the speed with which promoters are reactivated when phosphate is added to cultures of phosphate-limited cells. Promoters can be maximally reactivated within 10 min of phosphate addition (Botella et al., 2011). These studies illustrate the additional insights into promoter behaviour that can be obtained by high-resolution temporal analysis using GFP fusion technology.

High temporal resolution of promoter activity formed part of the analysis of the global network reorganization that occurs during adaptation of *B. subtilis* to different carbon sources (Buescher et al., 2012). Malate was added to *B. subtilis* cultures growing on glucose as sole carbon source (glycolytic metabolism), while glucose was added to cultures growing on malate as sole carbon source (gluconeogenic metabolism). The reorganization of the regulatory network that ensued was established using transcriptome, promoter activity (~300 transcriptional fusions), proteome and metabolome data, combined with statistical- and model-based analyses. The analysis showed that adaptation of glucose-grown cells to the addition of malate occurred rapidly and mainly at a post-transcriptional level. In contrast, the adaptation of malate-grown cells to the addition of glucose was slow and occurred mainly at a transcriptional level, involving almost half the genes in the regulatory network (Buescher et al., 2012).

While luciferase reporters have not been used for high-throughput analysis of gene expression in *B. subtilis*, several studies have demonstrated their applicability to, and utility in, this bacterium. The Dubnau group has adapted the firefly luciferase (*luc* from plasmid pGL3; Promega Corp. Wi, USA) for use in *B. subtilis*, generating transcriptional fusions to examine the expression of the *rapH* and *phrH* genes. RapH is a response regulator aspartate phosphatase whose activity is modulated by the PhrH protein (Mirouze et al., 2011a,b, 2012). They also generated *spo0A* and *comK* promoter fusions with the firefly luciferase *luc* gene to establish the expression of these promoters during growth and the onset of competence development and sporulation (Mirouze et al., 2011a,b, 2012). Promoter activity was established at very high temporal resolution (1.5-min intervals) providing insight into the kinetics of promoter activity in unprecedented detail, and with unexpected results. For example, *spo0A* transcription fluctuates during the developmental transition period in a manner that correlates with pauses in growth rate that may be caused by the release of RNA polymerase from ribosomal RNA transcripts during this period. It is proposed that these bursts of *spo0A* transcription may provide the temporal gate for entry of cells into the competent state (Mirouze et al., 2011a,b, 2012) and explain why competence development has a bimodal distribution in cell populations. This temporal gating mechanism differs fundamentally from the involvement of Spo0A in sporulation development where the cellular level of phosphorylated Spo0A is the determining factor. The cellular ATP level is an important consideration when using firefly luciferase, as outlined in the supporting material of the Mirouze et al. (2011a) study. The K_M value of ATP binding to the light generating binding site of firefly luciferase is 100 μM (DeLuca and McElroy, 1984). Thus, when comparing promoter activities under different experimental conditions, it is important that cellular ATP levels do not deviate to the extent that they influence light production. The Losick group have adapted and optimized the *luxABCDE* operon from *P. luminescens* for use in

B. subtilis to establish expression of *sinI* at high temporal resolution (Schmalisch et al., 2010; McLoon et al., 2011). A particular novelty of this study is that the light emitted by the bacterial luciferase was measured in real time in growing colonies and in biofilm pellicles during their formation and maturation (McLoon et al., 2011).

In summary, these studies illustrate the versatility of both the GFP and luciferase reporters showing how the establishment of promoter activity at high resolution can reveal the dynamics of genetic control mechanisms in unprecedented detail and establish novel features of regulation and regulatory networks.

7 METHODOLOGY FOR HIGH-THROUGHPUT ANALYSIS OF PROMOTER ACTIVITY WITH FINE TEMPORAL RESOLUTION IN *B. subtilis*

The following sections discuss important practical aspects associated with the determination of global promoter activity at high temporal resolution in *B. subtilis*. The available plasmids, procedures and methodologies are outlined, with emphasis on their advantages and disadvantages.

7.1 Promoter fragment selection and generation

With more than 12,000 prokaryotic genomes now partially or completely sequenced, most studies would use a directed approach to identify and clone promoter-containing DNA fragments. In the absence of detailed transcriptome knowledge, the majority of promoters can be assumed to be located within intergenic regions. However, it should be noted that the number of promoters was higher than predicted in a recent transcriptomic study in *B. subtilis* using high-density microarrays, with a significant number located 3' to the operon of interest generating antisense RNA (Nicolas et al., 2012). Delimiting the promoter-containing fragment to be cloned depends on the type of fusion to be made and its placement within the cell: that is, whether it is to be cloned into a replicating plasmid or integrated into the chromosome by a single-crossover event at the homologous locus or a double-crossover event at a heterologous neutral locus.

The ease with which constructs can be integrated into the chromosome has made this the method of choice in *B. subtilis*. In a recent study, promoter fusions were integrated into the chromosome at homologous loci by a single-crossover event (Botella et al., 2010, 2011). Promoter-containing fragments spanning each intergenic region were amplified by PCR (Botella et al., 2010, 2011; Buescher et al., 2012). The 3'-end of each PCR-generated fragment was located 18 bp upstream of the start codon of the first gene of the operon to avoid the ribosome binding site and to prevent formation of translational fusions. For intergenic regions of less than 400 bp, a fragment of 400 bp was chosen despite the 5'-end containing the 3'-end of the previous operon. There are several reasons for these choices. Integration directed by such a DNA fragment is non-mutagenic, generating a transcriptional promoter fusion while leaving the normal locus and its control regions intact. Thus expression of each promoter fusion is established in the wild-type genetic background. A high frequency of integrants is ensured with a minimum of 400 bp of DNA homology. When integrating promoter fusions at a neutral heterologous locus (e.g. *amyE* in *B. subtilis*) by a double-crossover event, the promoter-containing DNA fragment can be of any length, although the restrictions delimiting the 3′-end still apply. A LIC system is particularly suited to high-throughput generation of promoter fusions using several of the plasmids described in the following section (Botella *et al.*, 2010; Bisicchia *et al.*, 2010; Figure 1.1). LIC compatible sequences are designed into the 5′-ends of the primers used to amplify the promoter-containing DNA fragments as shown in Figure 1.1. Treatment of the fragments with T4 DNA polymerase in the presence of dTTP generates single-stranded ends that are compatible with similarly treated ends of plasmid pBaSysBioII (see below). This cloning methodology has several advantages including its suitability for automated high-throughout cloning procedures, a high transformation frequency, no background (false-positive) transformants and directional cloning of the insert so that the promoter always reads into the reporter gene.

7.2 Plasmids

The following briefly describes some plasmids used to generate transcriptional fusions that allow comparative analysis of promoter activities. Several plasmids designed specifically for use in *B. subtilis* are available from the *Bacillus* Genetic Stock Centre (the BGSC ID number is given in parenthesis).

7.2.1 pBaSysBioII

Plasmid pBaSysBioII (ECE223) is designed to integrate promoter fusions into the *B. subtilis* chromosome by a single-crossover event at homologous loci (Figure 1.1). Plasmid pBaSysBio II encodes a pBR322 replicon and ampicillin resistance gene allowing constructs to be cloned in *E. coli* (Botella *et al.*, 2010). It also encodes a promoterless *gfpmut3* gene isolated from pJBA27 (Bo Andersen *et al.*, 1998), with an optimized ribosome binding site (AAGGAGG) located 7 bp upstream of the ATG start codon, and a LIC cloning site. Digestion of the plasmid with *Sma*I, followed by treatment with T4 DNA polymerase in the presence of dATP, generates single-stranded ends that are compatible with the promoter-containing fragments with LIC ends that have been treated with T4 DNA polymerase in the presence of dTTP (Figure 1.1). Annealed plasmid and insert can be transformed directly into *E. coli* without ligation.

7.2.2 pGFPamy, pCFPamy, pYFPamy/pGFPbglS, pCFPbglS, pYFPbglS vector suite

A series of plasmids has been constructed to expand the methodology for generating promoter fusions (Bisicchia *et al.*, 2010). The *gfpmut3* (from pBaSysBioII), *yfp* (from pIYFP, Veening *et al.*, 2004) and *cfp* (from pDR200, Doan *et al.*, 2005)

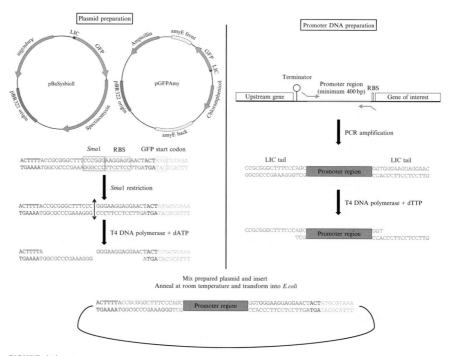

FIGURE 1.1

Promoter fragment and plasmid preparation for ligation-independent cloning. Plasmids pBaSysBioII and pGFPAmy both encode the *gfpmut3* reporter gene (GFP), a ribosome binding site (RBS) optimized for use in *B. subtilis* and a ligation-independent cloning (LIC) site located upstream of the RBS. Cleavage of either plasmid with *Sma*I followed by treatment with T4 DNA polymerase in the presence of dATP yields a linear plasmid with non-cohesive single-stranded ends. Promoter-containing fragments are generated by PCR using oligonucleotides (orange arrow) with a LIC tail (bent blue tail). Treatment of the promoter-containing fragment with T4 polymerase in the presence of dTTP yields linear fragments with non-cohesive single-stranded ends. The single-stranded regions of the linearized plasmid and promoter-containing DNA fragment are complementary so that upon annealing, a circular plasmid molecule is generated with directional insertion of the promoter so that it reads into the reporter *gfp* gene. This annealed plasmid transforms *E. coli* at high frequency despite being nicked. Moreover, no false positive clones are obtained, as the single-strand ends of the plasmid are non-cohesive. (For interpretation of the references to colour in this figure legend, the reader is referred to the online version of this chapter.)

reporter genes were cloned into plasmids that direct chromosomal integration by a double-crossover event at the *amyE* (plasmids pGFPamy, pCFPamy, pYFPamy) or the *bglS* (pGFPbglS, pCFPbglS, pYFPbglS) loci (Bisicchia et al., 2010). Each reporter gene has a ribosome binding site optimized for translation in *B. subtilis*, upstream of which is located a LIC cloning site (Figure 1.1). Integrants with the desired promoter fusions located at the *amyE* or *bglS* loci can be screened by a plate assay (Bisicchia et al., 2010). Briefly, transformants are stabbed onto plates containing 1% starch (for integrations at *amyE*) or 0.4% lichenan (for integrations at *bglS*) and onto replica plates without any addition and grown overnight. These plates are then flooded with iodine (0.1% in 1N HCl) or Congo red dye (0.4%). Colonies with reduced or without halos will have an integration at these loci. Stocks should be made from the corresponding colony on the replica plate. The half-life of both CFP and YFP in *B. subtilis* is approximately 2 h, enabling expression profiles of promoter fusions with these two reporter proteins to be compared (Bisicchia et al., 2010). This vector suite has several advantages and offers considerable versatility for fusion generation: (1) fusions are present in single copy on the chromosome; (2) the activity of promoters from other bacteria can be investigated in *B. subtilis* and (3) CFP and YFP are spectrally distinct so that the activities of two different promoters can be examined in the same strain. This is further facilitated by the use of two different integration sites (*amyE* and *bglS*) and selection for resistance to two different antibiotics (kanamycin and chloramphenicol). This suite of vectors is available at the *Bacillus* Genetic Stock Centre (http://www.bgsc.org/): ECE224, pGFP(*bglS*); ECE225, pCFP(*bglS*); ECE226, pYFP(*bglS*); ECE227, pGFP(*amyE*); ECE228, pCFP(*amyE*) and ECE229, pYFP(*amyE*).

7.2.3 pFSB79
This vector is designed to generate transcriptional *gfp* fusions. The *gfp*$^+$ gene is utilized on this plasmid. It encodes a fast-folding GFP variant with enhanced fluorescence (Scholz et al., 2000) although the stability of this protein in *B. subtilis* has not been established. Promoter fusions are integrated into the *amyE* locus of the *B. subtilis* chromosome with selection for chloramphenicol (Brigulla, et al., 2003).

7.2.4 pAH321
This vector is designed to generate transcriptional fusions using the *luxABCDE* system from *P. luminescens* (Schmalisch et al., 2010). Promoter-containing fragments must have *Eco*RI–*Sal*I ends for directional cloning into the *Eco*RI–*Sal*I-digested pAH321 plasmid (McLoon et al., 2011). The ribosome binding sites for all the *lux* genes have been optimized for expression in *B. subtilis* and the linearized recombinant plasmid is integrated into the neutral *sac* locus of the chromosome with selection for chloramphenicol resistance (McLoon et al., 2011).

7.3 High-throughput cloning and plasmid preparation
Promoter fusions can easily be generated manually in batches of up to 24 constructs or robotically using high-throughput procedures. The LIC procedure outlined above lends itself well to automation and high-throughput application because sets of forward primers have common 5′ extensions, as do the sets of reverse primers (Au et al., 2006; Fogg and Wilkinson, 2008). A list of target promoter sequences is established and compatible sets of PCR primer sequences with user-specified properties are generated by computer. Oligonucleotides are obtained from commercial suppliers in

lyophilized form in pairs of 96-well blocks, the first containing the forward primers, the second containing the reverse primers. Handling steps are performed with a TECAN robot (Tecan Group Ltd., Switzerland). Oligonucleotides are resuspended and diluted to 20 pmol/µl. PCRs to amplify DNA fragments are set up in 96-well trays with well volumes of 0.2 ml. Each reaction contains 1 µl each of the forward and reverse primers, 48 µl of a master mix (containing dNTPs, $MgSO_4$, $10 \times$ buffer and DNA polymerase) and 25 ng of chromosomal DNA. Having confirmed successful amplification of DNA fragments of the correct size (~ 400 bp) by gel electrophoresis, DNA products are purified using the QIAGEN 96-well PCR clean up kit (QIAGEN, Crawley, UK) with the steps carried out on the TECAN robot additionally incorporating a vacuum manifold. The concentration of the purified DNA is determined using a nanodrop spectrophotometer. LIC single-stranded DNA overhangs (Figure 1.1) are generated in a 96-well tray, adding to each well 0.2 pmol of each PCR product, 2.5 units of LIC qualified T4 DNA polymerase enzyme (Novagen), 2 µl of $10 \times$ T4 polymerase buffer (Novagen, Darmstadt, Germany), 1 µl of 100 mM dithiothreitol, and 2 µl of 25 mM dTTP in a final volume of 20 µl. The plate is incubated at 22 °C for 30 min and the reaction stopped by heat treatment at 75 °C for 20 min. Approximately 5 ng of each T4 DNA polymerase-treated PCR product is mixed with ~ 15 ng of linear pBaSysBioII with compatible LIC ends (Botella et al., 2010). Plasmid and insert are annealed for a minimum of 10 min and transformed into appropriate E. coli strains. Twenty microlitres of the transformation cell suspension from each of the 96 wells is plated onto 1 ml of solid Luria Bertani (LB) agar medium with appropriate antibiotics, that is, four plates each with 24 wells. Two colonies from each agar well are picked into 1 ml of liquid LB medium in 96-well plates (well volume 2 ml) and grown at 37 °C.

Plasmid DNA is isolated using a QIAGEN 96-well miniprep kit, in steps carried out on the TECAN robot using the vacuum manifold, as specified in supplier's instructions. The concentration of the purified plasmid is determined by nanospectrophotometry and inserts are verified by DNA sequencing. The success rate for cloning inserts with correct sequences was typically 80% for a 96-well plate. Plasmids encoding transcriptional fusions are transformed into B. subtilis by the procedure of Anagnostopoulos and Spizizen (1961).

7.4 Strain storage

The capability to generate large numbers of promoter fusion-containing strains requires the establishment of a coordinated system to store, transfer, and grow them and to determine promoter activity in a high-throughput manner. We have developed a methodology to perform these tasks that gives highly reproducible growth and expression profiles. We use a Boekel 96-pin replicator (Boekel Scientific, PA, USA) and a multichannel pipette for rapid manipulation and growth of strains in a 96-well format (Figure 1.2). Individual promoter fusion containing B. subtilis transformants are picked using toothpicks and inoculated into 0.75 ml LB broth in 96-well plates with a well capacity of 2 ml. Strains are not stored in the outer wells of the

7 Methodology for high-throughput analysis of promoter activity

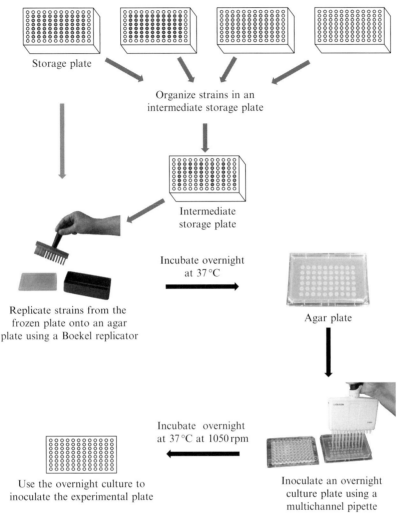

FIGURE 1.2

Multiplexed manipulation of promoter fusion-containing strains for growth and expression studies. Promoter fusion-containing strains of B. subtilis are individually picked from transformation plates into a 96-well storage plate with a well capacity of 2 ml (plates with blue, red, green, yellow spots). The outer wells of the plate are not utilized. Strains are grown overnight at 37 °C in these plates, glycerol is added and the plates are stored at −70 °C. To initiate expression analysis of the strains from a particular storage plate, a Boekel 96 pin replicator is used to transfer an inoculum to agar medium on a square petri plate (blue arrow), which is then grown overnight at 37 °C. Should it be required to examine expression of strains distributed among several storage plates, best results are obtained by generating an intermediate storage plate from which strains are then transferred to a square petri plate using the Boekel replicator (red arrows). Overnight cultures of each strain are generated by storage plate, as growth and promoter fusion expression in these wells in the experimental plate is unreliable due to evaporation. Cultures are grown overnight with shaking after which 0.3 ml sterile 50% glycerol is added and mixed by pipetting. The plate is covered with a lid, sealed with parafilm and stored at $-70\ °C$. The plate is named and strains identified using a grid (Figure 1.3). Strains should be placed in these storage plates in the order in which their expression will be established, as all subsequent manipulations are performed with the Boekel replicator or multichannel pipette (blue arrow, Figure 1.3). Should it be required to determine expression of selected fusions distributed among many storage plates, we routinely pick (with toothpicks) these strains from the original storage plate into a new storage plate, organizing them as required for the expression studies and preparing it for storage as previously described (red arrows, Figure 1.3).

As transfer of strains from storage plates to the experimental plate is multiplexed, the strains and their location on the storage plate should be carefully planned. For example, each storage plate should have the required control strains, namely, a strain without a promoter fusion for every genetic background represented on the plate. This is necessary so that background autofluorescence can be subtracted at each timepoint of the growth curve for each strain. Thus strains with the same genetic background should generally be stored together. It is prudent also to consider the parameters and variables to be tested and compared so that the plates can be designed to contain the relevant fusion-containing strains, thereby avoiding the necessity of having to compare expression profiles of strains grown on different plates under different culture conditions. Thus, for example, strains with promoter fusions belonging to a single regulon or stimulon would typically be stored on the same plate, together with the required control strains. Nevertheless, our experience is that this methodology yields very reproducible growth and expression profiles, even when strains are grown on different 96-well plates.

Three independent transformants of each promoter fusion construct are placed on the plate for growth and expression analyses (Figure 1.3). In some cases, the expression profile of all three will be similar but the absolute level of GFP expression in one strain may be an exact multiple (usually twofold, rarely threefold) of the other two strains at all time points. This is assumed to result from there being multiple copies of the promoter fusion on the chromosome, generated either by amplification on the chromosome or through integration of a multimeric plasmid. However, for fusions of exceptional experimental significance, we routinely generate promoter fusions

---

inoculating a 96-well plate (well capacity 0.2 ml) from the freshly grown colonies using a multichannel pipette and incubating overnight at 37 °C with shaking. An experimental plate, in which growth and GFP expression will be monitored, is prepared by transferring 1 μl of the overnight culture to a similar experimental plate using a multichannel pipette. The outer wells are filled with culture medium but are not inoculated, as growth and expression of cultures in these wells are less reproducible due to evaporation. (For interpretation of the references to colour in this figure legend, the reader is referred to the online version of this chapter.)

	1	2	3	4	5	6	7	8	9	10	11	12
A												
B	Media Blank	background strain	S 1.1	S 2.1	S 3.1	S 4.1	S 5.1	S 6.1	S 7.1	S 8.1	S 9.1	
C	Media Blank	background strain	S 1.2	S 2.2	S 3.2	S 4.2	S 5.2	S 6.2	S 7.2	S 8.2	S 9.2	
D	Media Blank	background strain	S 1.3	S 2.3	S 3.3	S 4.3	S 5.3	S 6.3	S 7.3	S 8.3	S 9.3	
E	Media Blank	S 10.1	S 11.1	S 12.1	S 13.1	S 14.1	S 15.1	S 16.1	S 17.1	S 18.1	background strain	
F	Media Blank	S 10.2	S 11.2	S 12.2	S 13.2	S 14.2	S 15.2	S 16.2	S 17.2	S 18.2	background strain	
G	Media Blank	S 10.3	S 11.3	S 12.3	S 13.3	S 14.3	S 15.3	S 16.3	S 17.3	S 18.3	background strain	
H												

	1	2	3	4	5	6	7	8	9	10	11	12
A												
B	Media Blank	C1	C2	C3	C4	C5	C6	C7	C8	C9	C10	
C	Media Blank	E1	E2	E3	E4	E5	E6	E7	E8	E9	E10	
D	Media Blank	C11	C12	C13	C14	C15	C16	C17	C18	C19	C20	
E	Media Blank	E1	E2	E3	E4	E5	E6	E7	E8	E9	E10	
F	Media Blank	C21	C22	C23	C24	C25	C26	C27	C28	C29	C30	
G	Media Blank	E21	E22	E23	E24	E25	E26	E27	E28	E29	E30	
H												

FIGURE 1.3

Two possible formats for distributing test strains in 96-well plates. The upper format is designed to examine 18 transcriptional fusions (S1–S18) in triplicate (S1.1, S1.2, S1.3, etc.) in a single genetic background. The positioning of six control strains (i.e. the strains used to determine background autofluorescence that do not contain promoter fusions) is shown (yellow). The lower format is designed to examine the expression of 30 strains in which the transcriptional fusions are each in a different genetic background (E1–E30), together with a control strain for each of these backgrounds (C1–C30). All wells should be filled with media, although the outer wells are not used for culture growth. (For colour version of this figure, the reader is referred to the online version of this chapter.)

integrated at the *amyE* locus to ensure the expression data are obtained from a single copy of the promoter fusion.

7.5 Growth and data collection

The following procedure was found to give very reproducible growth and expression profiles. Fresh colonies of strains from the desired storage plate(s) are prepared by transferring an inoculum onto agar in square plates using a 96-well Boekel replicator and grown overnight (Figure 1.2). The resultant agar plate has a grid of well-separated freshly grown colonies of each strain (Figure 1.2). An overnight broth culture of each strain is generated by inoculating a 96-well plate (0.5 ml well capacity) with material from each of the colonies on the agar plate using a multichannel pipette (Figure 1.2). Disposable tips on a multichannel pipette are touched to the colonies and then placed into media in a 96-well plate with some agitation. Each well contains 100 μl of broth and antibiotics as appropriate. This plate is incubated in a sealed plastic bag (to prevent evaporation) overnight at 37 °C with shaking. The experimental plate is prepared by transferring 1 μl of each overnight culture using the multichannel pipette to the wells of the experimental plate containing 100 μl of the desired culture medium. All wells of the experimental plate should contain culture medium even though the outer wells are not inoculated with any strains. These wells are used for background fluorescence readings, as controls to indicate that cross-well contamination has not occurred and to maintain humidity and reduce evaporation in the inner wells.

A typical protocol for an experiment starting with frozen stocks in the 96-well storage plate is as follows:

1. Remove the storage plate from the −70 °C freezer and allow to thaw for ~5 min.
2. Generate a grid of fresh colonies on a square culture dish containing LB agar with appropriate antibiotic additions, using a sterilized Boekel 96-pin replicator.
3. Generate overnight cultures (100 μl) by inoculating a 96-well plate (Falcon 353072) from the grid of strains using a multichannel pipette with disposable tips. This 96-well plate is placed in a plastic bag to prevent evaporation and cultivated overnight at 37 °C under constant shaking of 1050 rpm in a Heidolph titramax 1000 (Heidolph UK, Essex, UK) incubator.
4. The experimental plate is prepared by transferring 1 μl of the overnight culture to the 96-well experimental plate (Greiner 655180) containing the appropriate medium (100 μl per well) using a multichannel pipette.
5. The experimental plate is then incubated at the desired temperature and aeration conditions in a Synergy II plate reader (BioTek Germany, Bad Friedrichshall, DE).
6. Injections can be performed to all or selected wells at the desired time intervals using the injectors of the Synergy II plate reader.
7. User-defined protocols will determine the time intervals at which optical density (OD_{600}) and reporter protein emission readings are to be recorded. We typically take readings every 10 min for the duration of the growth experiment, although intervals as short as 1 min have also been successful (Botella et al., 2010).
8. Growth (OD_{600}) and expression data are stored, then exported to an Excel spreadsheet file for subsequent analysis.

7.6 Data analysis

Growth of the individual cultures in the 96-well plate is monitored by measuring optical density at 600 nm (OD_{600}), while fluorescence monitors expression of the GFP protein. Because the length of the light path in a 96-well plate is shorter than the 1-cm path-length in a standard spectrophotometer cuvette, the path-length is corrected by taking an OD_{900} and OD_{977} reading for each well at the beginning of the experiment. The path length is corrected by multiplying each OD_{600} reading (after subtraction of blank values) by [0.18/(OD$_{977}$ − OD$_{900}$)]. This calculation can be approximated by multiplying each OD$_{600}$ reading by 5 when a 100-μl culture is grown in a 96-well plate. The excitation and emission settings for GFPmut3 are 485/20 and 528/20 nm, respectively. For YFP, the excitation and emission settings are 500/27 and 540/25 nm, respectively, while for CFP, the settings are 420/50 and 485/20 nm, respectively.

B. subtilis strains without GFP fusions have background autofluorescence, which must be subtracted from the GFP emission values of promoter fusion-containing strains before calculating promoter activities. The level of background autofluorescence is dependent on the size and shape of the cells and on the density of the culture. To perform this correction, OD$_{600}$ and fluorescence data are collected for six cultures of control strains (i.e. those without promoter fusions). A curve is plotted with the OD$_{600}$ values on the abscissa and fluorescence values on the ordinate axes. A polynomial trend line (order 4) is fit to the resulting curve and its equation is determined. This equation is used to subtract background autofluorescence from the measured fluorescence of each fusion-containing strain at the same optical density. Because strains with different genetic backgrounds can have different size, shape or growth characteristics, and hence levels of autofluorescence, controls for each genetic background should be grown on every plate. This polynomial-mediated subtraction should also be performed on the five remaining control (non-fusion-containing) strains to assess the efficacy of background subtraction. If necessary background subtraction can be further refined by subtracting the maximum autofluorescence values that remain after the polynomial treatment of the control strains.

Some difficulties in accurately subtracting the background fluorescence can arise if the optical density of the culture remains constant for an extended period of time or if the optical density decreases. In these situations, as the same OD$_{600}$ value is being obtained several times along the growth curve, more than one fluorescent value can be assigned to the same optical density point.

Promoter activity is then calculated at each time point by dividing the derivative of the background-corrected fluorescence by the OD$_{600}$ [(dGFP/dt)/OD$_{600}$]. This gives the amount of GFP synthesized in the designated time period normalized to the cell density of the culture. As GFP is very stable in *B. subtilis*, this reading is a real-time measurement of the activity of the promoter under study in each time interval. Promoter activity curves can be smoothed by plotting the average of three successive time points.

7.7 Visualization of promoter activity

The activity of hundreds of promoters can be established with relative ease for a variety of experimental conditions using the high-throughput procedures described (Botella *et al.*, 2010, 2011). We have used heat maps to visualize and analyze expression trends in these large data sets. Heat Map Generator (HMG, http://bioinf.gen.tcd.ie/HMG) provides a user-friendly web interface to quickly create publication quality heat map images. Extra functionality is included to facilitate the pre-processing of the data and the formatting of images. The main input consists of a promoter activity profile in the form of a tab-delimited text file. To perform background subtraction, an additional file with OD values must be uploaded. The first pre-processing step involves data smoothing by user-defined averaging of the values from neighbouring time-points. Any negative values can be set to zero. If the dataset contains experimental or technical replicates, HMG can merge these using either mean or median values. In the latter case, the user can also opt to keep the values of the clone that is in general closest to the median, that is, a preferred clone is determined automatically by HMG. The final processing step includes the scaling of all values to a range from (0–1). The intermediate output after each data transformation step can be inspected and downloaded. The processed data are then visualized as an online HTML table where the cells are coloured according to the promoter activity values. This table is presented through the online-spreadsheet part of ArrayPipe (Hokamp et al., 2004), a microarray data analysis tool. It allows colours to be changed as well as the filtering and sorting of columns. This is particularly useful after a cluster analysis, which is available within ArrayPipe as an integrated module. Users can also upload annotation information for attachment to the table. Once the output is in the desired format, a heatmap can be generated in PNG or scalable PDF file formats (Figure 1.4). The web-page for generating the image can be password-protected, bookmarked and shared with others. Detailed user documentation is available online.

8 CONCLUSIONS

The systems approach attempts to describe and understand biological systems in a global and interactive manner. This approach has emerged from new technological capabilities such as next-generation sequencing, microarrays, RNAseq and high-throughput analysis of the promoter fusions (live cell arrays) to establish transcriptomes, promoter activity profiles, proteomes, interactomes and metabolomes. Here, we have described the capabilities and uses of gene fusion technology to determine promoter activity with high temporal resolution on a global scale. The functionality, reproducibility and low cost of this technology means it can be widely adapted and applied. The resultant global and dynamic view has revealed several novel regulatory features of gene expression in *E. coli* and *B. subtilis*. When such analyses are integrated with other global analyses, a very detailed description of the behaviour of a biological system can be established (Buescher et al., 2012). Clearly, this technology will make a seminal contribution towards formulating dynamic gene expression models and the development of a more global view of how biological systems function.

Acknowledgements

Research in the Devine laboratory is supported by Science Foundation Ireland Principal Investigator Award 08/IN.1/B1859. Development work on live cell arrays

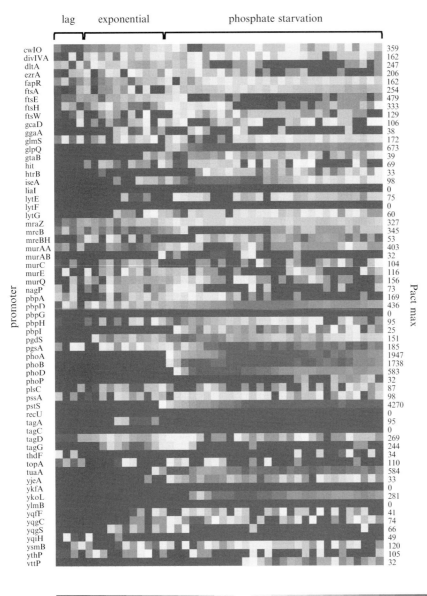

FIGURE 1.4

Expression profiles of promoter fusions visualized in heat map format. The expression profiles of 63 randomly selected promoter *gfpmut3* fusions generated by Heat Map Generator (http://bioinf.gen.tcd.ie/HMG). The values of each profile are presented on a 0–1 scale, normalized to the maximum promoter activity value (P_{act} max) of each fusion which is given at the right hand side of the map. The colour legend signifies the percentage of P_{act} max represented by each colour. By combining the colour values with the P_{act} max values, the strengths of individual promoters at different points of the growth cycle (e.g. lag, exponential growth, phosphate limitation) can be compared. (For interpretation of the references to colour in this figure legend, the reader is referred to the online version of this chapter.)

in *B. subtilis* in the Devine and Wilkinson laboratories was supported by the EU-funded BaSysBio project (LSHG-CT-2006-037469). Karsten Hokamp is supported by Science Foundation Ireland.

References

Anagnostopoulos, C. and Spizizen, J. (1961). Requirements for transformation in *Bacillus subtilis*. *J. Bacteriol.* **81**, 741–746.

Andersen, J. B., Sternberg, C., Poulsen, L. K., Bjørn, S. P., Givskov, M., and Molin, S. (1998). New unstable variants of green fluorescent protein for studies of transient gene expression in bacteria. *Appl. Environ. Microbiol.* **64**(6), 2240–2246.

Au, K., Berrow, N. S., Blagova, E., Boucher, I. W., Boyle, M. P., Brannigan, J. A., Carter, L. G., Dierks, T., Folkers, G., Grenha, R., Harlos, K., Kaptein, R., Kalliomaa, A. K., Levdikov, V. M., Meier, C., Milioti, N., Moroz, O., Muller, A., Owens, R. J., Rzechorzek, N., Sainsbury, S., Stuart, D. I., Walter, T. S., Waterman, D. G., Wilkinson, A. J., Wilson, K. S., Zaccai, N., Esnouf, R. M., and Fogg, M. J. (2006). Application of high-throughput technologies to a structural proteomics-type analysis of *Bacillus anthracis*. *Acta Crystallogr. D* **62**, 1267–1275.

Bisicchia, P., Botella, E., and Devine, K. M. (2010). Suite of novel vectors for ectopic insertion of GFP, CFP and IYFP transcriptional fusions in single copy at the *amyE* and *bglS* loci in *Bacillus subtilis*. *Plasmid* **64**, 143–149.

Bjarnason, J., Southward, C. M., and Surette, M. G. (2003). Genomic profiling of iron-responsive genes in *Salmonella enterica* serovar *typhimurium* by high-throughput screening of a random promoter library. *J. Bacteriol.* **185**, 4973–4982.

Botella, E., Fogg, M., Jules, M., Piersma, S., Doherty, J., Hansen, A., Denham, E. L., Le Chat, L., Veiga, P., Bailey, K., Lewis, P. J., van Dijl, J. M., Aymerich, S., Wilkinson, A. J., and Devine, K. M. (2010). pBaSysBioII: an integrative plasmid generating *gfp* transcriptional fusions for high-throughput analysis of gene expression in *Bacillus subtilis*. *Microbiology* **156**, 1600–1608.

Botella, E., Hübner, S., Hokamp, K., Hansen, A., Bisicchia, P., Noone, D., Powell, L., Salzberg, L. I., and Devine, K. M. (2011). Cell envelope gene expression in phosphate-limited *Bacillus subtilis* cells. *Microbiology* **157**, 2470–2484.

Brigulla, M., Hoffmann, T., Krisp, A., Völker, A., Bremer, E., and Völker, U. (2003). Chill induction of the SigB-dependent general stress response in *Bacillus subtilis* and its contribution to low-temperature adaptation. *J. Bacteriol.* **185**, 4305–4314.

Buescher, J. M., Liebermeister, W., Jules, M., Uhr, M., Muntel, J., Botella, E., Hessling, B., Kleijn, R. J., Le Chat, L., Lecointe, F., Mäder, U., Nicolas, P., Piersma, S., Rügheimer, F., Becher, D., Bessieres, P., Bidnenko, E., Denham, E. L., Dervyn, E., Devine, K. M., Doherty, G., Drulhe, S., Felicori, L., Fogg, M. J., Goelzer, A., Hansen, A., Harwood, C. R., Hecker, M., Hubner, S., Hultschig, C., Jarmer, H., Klipp, E., Leduc, A., Lewis, P., Molina, F., Noirot, P., Peres, S., Pigeonneau, N., Pohl, S., Rasmussen, S., Rinn, B., Schaffer, M., Schnidder, J., Schwikowski, B., van Dijl, J. M., Veiga, P., Walsh, S., Wilkinson, A. J., Stelling, J., Aymerich, S., and Sauer, U. (2012). Global network reorganization during dynamic adaptations of *Bacillus subtilis* metabolism. *Science* **335**, 1099–1103.

Champe, S. P. and Benzer, S. (1962). A active cistron fragment. *J. Mol. Biol.* **4**, 288–292.

Chechik, G. and Koller, D. (2009). Timing of gene expression responses to environmental changes. *J. Comput. Biol.* **16**, 279–290.

Chechik, G., Oh, E., Rando, O., Weissman, J., Regev, A., and Koller, D. (2008). Activity motifs reveal principles of timing in transcriptional control of the yeast metabolic network. *Nat. Biotechnol.* **26**, 1251–1259.

Cormack, B. P., Valdivia, R. H., and Falkow, S. (1996). FACS-optimized mutants of the green fluorescent protein (GFP). *Gene* **173**, 33–38.

Courcelle, J., Khodursky, A., Peter, B., Brown, P. O., and Hanawalt, P. C. (2001). Comparative gene expression profiles following UV exposure in wild-type and SOS-deficient *Escherichia coli*. *Genetics* **158**, 41–64.

DeLuca, M. and McElroy, W. D. (1984). Two kinetically distinguishable ATP sites in firefly luciferase. *Biochem. Biophys. Res. Commun.* **123**, 764–770.

Doan, T., Marquis, K. A., and Rudner, D. Z. (2005). Subcellular localization of a sporulation membrane protein is achieved through a network of interactions along and across the septum. *Mol. Microbiol.* **55**, 1767–1781.

Fogg, M. J. and Wilkinson, A. J. (2008). Higher-throughput approaches to crystallization and crystal structure determination. *Biochem. Soc. Trans.* **36**, 771–775.

Fraga, H. (2008). Firefly luminescence: a historical perspective and recent developments. *Photochem. Photobiol. Sci.* **7**, 146–158.

Goh, E. B., Yim, G., Tsui, W., McClure, J., Surette, M. G., and Davies, J. (2002). Transcriptional modulation of bacterial gene expression by subinhibitory concentrations of antibiotics. *Proc. Natl. Acad. Sci. USA* **99**, 17025–17030.

Hokamp, K., Roche, F. M., Acab, M., Rousseau, M. E., Kuo, B., Goode, D., Aeschliman, D., Bryan, J., Babiuk, L. A., Hancock, R. E., and Brinkman, F. S. (2004). ArrayPipe: a flexible processing pipeline for microarray data. *Nucleic Acids Res.* **32**(Web Server issue), W457–W459.

Janniere, L., Niaudet, B., Pierre, E., and Ehrlich, S. D. (1985). Stable gene amplification in the chromosome of *Bacillus subtilis*. *Gene* **40**, 47–55.

Kalir, S., McClure, J., Pabbaraju, K., Southward, C., Ronen, M., Leibler, S., Surette, M. G., and Alon, U. (2001). Ordering genes in a flagella pathway by analysis of expression kinetics from living bacteria. *Science* **292**, 2080–2083.

McLoon, A. L., Guttenplan, S. B., Kearns, D. B., Kolter, R., and Losick, R. (2011). Tracing the domestication of a biofilm-forming bacterium. *J. Bacteriol.* **193**, 2027–2034.

Mirouze, N., Prepiak, P., and Dubnau, D. (2011a). Fluctuations in *spo0A* transcription control rare developmental transitions in *Bacillus subtilis*. *PLoS Genet.* **7**, e1002048.

Mirouze, N., Parashar, V., Baker, M. D., Dubnau, D. A., and Neiditch, M. B. (2011b). An atypical Phr peptide regulates the developmental switch protein RapH. *J. Bacteriol.* **193**, 6197–6206.

Mirouze, N., Desai, Y., Raj, A., and Dubnau, D. (2012). Spo0A P imposes a temporal gate for the bimodal expression of competence in *Bacillus subtilis*. *PLoS Genet.* **8**, e1002586.

Nicolas, P., Mäder, U., Dervyn, E., Rochat, T., Leduc, A., Pigeonneau, N., Bidnenko, E., Marchadier, E., Hoebeke, M., Aymerich, S., Becher, D., Bisicchia, P., Botella, E., Delumeau, O., Doherty, G., Denham, E. L., Fogg, M. J., Fromion, V., Goelzer, A., Hansen, A., Härtig, E., Harwood, C. R., Homuth, G., Jarmer, H., Jules, M., Klipp, E., Le Chat, L., Lecointe, F., Lewis, P., Liebermeister, W., March, A., Mars, R. A., Nannapaneni, P., Noone, D., Pohl, S., Rinn, B., Rügheimer, F., Sappa, P. K., Samson, F., Schaffer, M., Schwikowski, B., Steil, L., Stülke, J., Wiegert, T., Devine, K. M., Wilkinson, A. J., van Dijl, J. M., Hecker, M., Völker, U.,

Bessières, P., and Noirot, P. (2012). Condition-dependent transcriptome reveals high-level regulatory architecture in *Bacillus subtilis*. *Science* **335**, 1103–1106.

Ralser, M., Wamelink, M. M., Latkolik, S., Jansen, E. E., Lehrach, H., and Jakobs, C. (2009). Metabolic reconfiguration precedes transcriptional regulation in the antioxidant response. *Nat. Biotechnol.* **27**, 604–605.

Ronen, M., Rosenberg, R., Shraiman, B. I., and Alon, U. (2002). Assigning numbers to the arrows: parameterizing a gene regulation network by using accurate expression kinetics. *Proc. Natl. Acad. Sci. USA* **99**, 10555–10560.

Schmalisch, M., Maiques, E., Nikolov, L., Camp, A. H., Chevreux, B., Muffler, A., Rodriguez, S., Perkins, J., and Losick, R. (2010). Small genes under sporulation control in the *Bacillus subtilis* genome. *J. Bacteriol.* **192**, 5402–5412.

Scholz, O., Thiel, A., Hillen, W., and Niederweis, M. (2000). Quantitative analysis of gene expression with an improved green fluorescent protein. *Eur. J. Biochem.* **267**, 1565–1570.

Silander, O. K., Nikolic, N., Zaslaver, A., Bren, A., Kikoin, I., Alon, U., and Ackermann, M. (2012). A genome-wide analysis of promoter-mediated phenotypic noise in *Escherichia coli*. *PLoS Genet.* **8**, e1002443.

Silhavy, T. J. and Beckwith, J. R. (1985). Uses of *lac* fusions for the study of biological problems. *Microbiol. Rev.* **49**, 398–418.

Slauch, J. M. and Silhavy, T. J. (1991). Genetic fusions as experimental tools. *Methods Enzymol.* **204**, 213–248.

Stepanenko, O. V., Stepanenko, O. V., Shcherbakova, D. M., Kuznetsova, I. M., Turoverov, K. K., and Verkhusha, V. V. (2011). Modern fluorescent proteins: from chromophore formation to novel intracellular applications. *Biotechniques* **51**, 313–327.

Tsien, R. Y. (1998). The green fluorescent protein. *Annu. Rev. Biochem.* **67**, 509–544.

Veening, J. W., Smits, W. K., Hamoen, L. W., Jongbloed, J. D., and Kuipers, O. P. (2004). Visualization of differential gene expression by improved cyan fluorescent protein and yellow fluorescent protein production in *Bacillus subtilis*. *Appl. Environ. Microbiol.* **70**, 6809–6815.

Waidmann, M. S., Bleichrodt, F. S., Laslo, T., and Riedel, C. U. (2011). Bacterial Luciferase Reporters: the swiss army knife of molecular biology. *Bioeng. Bugs.* **2**(1), 8–16.

Wilson, T. and Hastings, J. W. (1998). Bioluminescence. *Annu. Rev. Cell Dev. Biol.* **14**, 197–230.

Young, M. (1984). Gene amplification in *Bacillus subtilis*. *J. Gen. Microbiol.* **130**, 1613–1621.

Zaslaver, A., Mayo, A. E., Rosenberg, R., Bashkin, P., Sberro, H., Tsalyuk, M., Surette, M. G., and Alon, U. (2004). Just-in-time transcription program in metabolic pathways. *Nat. Genet.* **36**, 486–491.

Zaslaver, A., Bren, A., Ronen, M., Itzkovitz, S., Kikoin, I., Shavit, S., Liebermeister, W., Surette, M. G., and Alon, U. (2006). A comprehensive library of fluorescent transcriptional reporters for *Escherichia coli*. *Nat. Methods* **3**, 623–628.

Zaslaver, A., Kaplan, S., Bren, A., Jinich, A., Mayo, A., Dekel, E., Alon, U., and Itzkovitz, S. (2009). Invariant distribution of promoter activities in *Escherichia coli*. *PLoS Comput. Biol.* **5**, e1000545.

Zhang, X., Wiseman, S., Yu, H., Liu, H., Giesy, J. P., and Hecker, M. (2011). Assessing the toxicity of naphthenic acids using a microbial genome wide live cell reporter array system. *Environ. Sci. Technol.* **45**, 1984–1991.

CHAPTER

Data mining for microbiologists

2

J.S. Hallinan[1]

*School of Computing Science and Centre for Bacterial Cell Biology, Newcastle University,
Newcastle upon Tyne, United Kingdom*
[1]*Corresponding author: e-mail address: j.s.hallinan@ncl.ac.uk*

1 INTRODUCTION

Microbiologists are drowning in data. Gigabytes of genomic, transcriptomic, proteomic, metabolic and interaction data are produced every day. Most of this data is eventually deposited in freely accessible databases, of which there are many: the 2012 *Nucleic Acids Research* Database issue reports on 1380 active online databases containing information about everything from the composition of DNA sequences to the details of protein complex formation (Galperin and Fernández-Suárez, 2012). Much of this data does not make it into the peer-reviewed literature. High-throughput experiments generate a considerable amount of data that is not of primary interest to the research group carrying out the experiments. For example, a microarray experiment can provide a snapshot of mRNA levels for every gene in a genome, but in general only those genes identified as being significantly up- or down-regulated are of interest in the research domain of the authors. The rest of the data are usually deposited in an appropriate database, but will not be published in the peer-reviewed literature.

For those interested in the information captured in unpublished datasets, exploring each database individually is prohibitively time consuming. Even if each database search takes only 5 min, finding all of the stored information about a single gene in those 1380 databases would take around 115 h of work. And that is without even considering the additional time required to read and assimilate the relevant literature.

Despite this embarrassment of information riches, a significant proportion of genes in most species are un-annotated, or annotated only on the basis of similarity to other genes. Probably the most well-studied microbe is the baker's yeast *Saccharomyces cerevisiae*, the first eukaryote to be sequenced (Goffeau *et al.*, 1996). *S. cerevisiae* has around 6200 open reading frames (ORFs). As of May 2012 the Munich Information on Protein Sequences (MIPS) database[1] lists 661 of these as

---

[1]http://mips.helmholtz-muenchen.de/genre/proj/yeast/.

producing a "protein of unknown function"; around 10% of the genome is a complete mystery. There is therefore considerable interest in using computational methods to address important problems such as the prediction of protein function and the identification of possible interactions between proteins.

The field of data mining was originally developed in the 1980s to make the most of business and marketing data, of the type that is produced every time a credit card transaction is made, or an electronic order is filled. Data mining is now widely applied to the investigation of large datasets in many other fields, including molecular biology. Indeed, it has been stated that "DNA sequencing and data mining have become almost as central to biology as transcription and translation are to life" (Yandell and Majoro, 2002).

This review aims to provide the working microbiologist with a guide to what can be achieved using data mining. It covers the basic principles of data mining, and describes the data mining life cycle. The algorithms most widely used for data mining in microbiology are described, together with an indication of the types of problems to which these algorithms have been applied. However, since data mining is a very wide field of research, it is not possible in this review to cover all of the relevant algorithms. Where appropriate, references are provided to more comprehensive textbooks, but for the most part we have limited the original literature to recent publications that indicate how data mining is currently being applied in microbiology.

2 WHAT IS DATA MINING?

A broad definition of data mining is "a set of mechanisms and techniques, realised in software, to extract hidden information from data" (Coenen, 2011). As Coenen points out, "hidden" is the key word in this definition; simply retrieving from GenBank a list of genes meeting a particular criterion is not data mining: the result of such a query is not information, but data. Data mining involves identifying patterns within and between data sources, and deducing their meaning. Data mining was originally based on conventional statistics but, as the field of computational intelligence (CI) developed over subsequent decades, many CI algorithms were eagerly seized upon by data miners. Today both statistical and CI approaches are used in a complementary manner.

Data mining algorithms were developed in parallel in several distinct fields, and consequently there is some confusion in terminology. Some forms of data mining are also known as Knowledge Discovery in Databases,[2] or pattern recognition (Kennedy, 1997), and there is considerable overlap between the fields of CI, Machine Learning and Artificial Intelligence.

---

[2]http://www.igi-global.com/journal/international-journal-knowledge-discovery-bioinformatics/1143.

Unlike most areas of microbiology research, data mining is not necessarily hypothesis driven, although hypothesis-driven algorithms, notably statistical approaches, are used. The aim of the data mining process is to find existing, but previously unidentified, patterns and interactions within large datasets, a process sometimes referred to as a "fishing expedition", or, more formally, *data-driven* research. Data-driven research must be undertaken with care. In large datasets apparently valid correlations, which appear to make biological sense, can arise by chance, particularly when large datasets are repeatedly analysed. A *p*-value of 0.001 is often regarded as indicating statistical significance. However, all that this number means is that the observed value is likely to arise by chance only once in 1000 trials. Data mining often involves thousands of tests performed on large, noisy datasets. Careful human scrutiny and analysis of the results is essential.

There are literally hundreds of algorithms which can and have been used for data mining in thousands of studies; a 2009 bibliometric study of the Web of Knowledge database identified nearly 10,000 journal articles on data mining, published between 1962 and 2008 (Shuang *et al.*, 2009). Amazon[3] currently lists over 20,000 books about data mining (not all of which will be relevant!). In the interest of not adding to this number, this review is limited to describing the algorithms most commonly used with microbiological data, with an indication of the ways in which the algorithms have been applied. Where possible, pointers are provided to freely available software implementing the various algorithms. Although there are also numerous commercial data mining products available, it can be valuable for interested researchers to have access to software that can be installed and investigated without a significant impact upon the budget of the relevant grant.

Algorithms suitable for data mining can be organised into a conceptual hierarchy (Figure 2.1).

Although, in the following sections, each algorithm is discussed individually, it is worth noting that many projects involve the use of multiple data mining algorithms, either sequentially or in parallel. The sequential application of algorithms results in the generation of workflows in which the outputs of one analysis become the inputs to the next. For example, data integration can be used to construct a network of interactions between proteins; the network can then be subjected to a clustering algorithm in order to identify functional modules; a Gene Ontology (GO; Ashburner *et al.*, 2000) over-representation analysis performed for each cluster; and then a protein function inference algorithm applied to the cluster members to predict the potential functions of un-annotated proteins. The application of different algorithms, in parallel to the same data, and designed to perform the same task, can provide additional insights into datasets. A Stepwise Linear Discriminant Analysis, for example, produces a classifier, and a ranked list of the most informative features contributing to the classifier. A Decision Tree does the same. Application of both algorithms to the same dataset, and comparison of the results, can confirm (or not!) the importance of specific features.

---

[3]http://www.amazon.co.uk/.

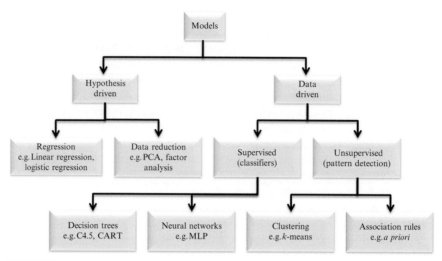

FIGURE 2.1

A categorisation of data mining algorithms.

3 THE DATA MINING PROCESS

Data mining involves the storage and handling of large amounts of data, potentially requiring significant amounts of time and computational effort. It is therefore important to approach the process in a structured manner (Figure 2.2).

The first step in the data mining process is the acquisition of domain knowledge. For a microbiologist this step is rarely a problem, but if the team includes people with data mining or computational expertise, but not necessarily an in-depth knowledge of the problem domain, it is essential that these team members acquire at least some domain knowledge. Insufficient domain knowledge can lead either to a failure to identify inconsistent data, or to the selection of inappropriate algorithms.

The objectives of the research, and the hypotheses upon which they are based, must be clearly stated at the beginning of the project. Communication of essential knowledge between team members from different disciplines can be a difficult and necessarily on-going process (Faris et al., 2011).

Appropriate data sources must then be identified and the data acquired. Data generated in-house should be securely stored, preferably in a structured database, and regularly backed up. Online databases usually offer a means to download part or all of the data they hold, but different databases use different formats. These formats range from text files intended primarily for human interpretation (such as GenBank), to highly structured formats intended primarily for computational access (such as any of the many flavours of the machine-readable extensible mark-up language, XML).

Because data mining is generally applied to very large datasets, the quality of the data is important. Most high-throughput datasets are noisy and incomplete, and contain unknown proportions of false positives (e.g. protein interactions which are

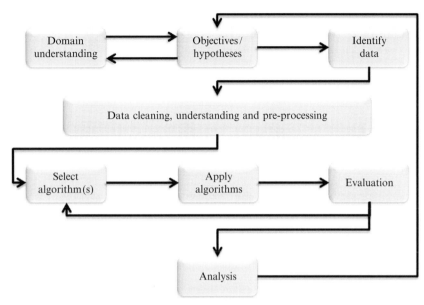

FIGURE 2.2

The data mining life cycle.

reported as existing, although they do not occur *in vivo*) and false negatives (interactions which are not recorded in the databases, but which do occur in nature). Most data mining algorithms attempt to take these issues into account, but the old adage "garbage in, garbage out" remains true. Data cleaning and pre-processing is perhaps the most important step of the entire data mining process, and can occupy up to 80% of the time taken for the project (Witten *et al.*, 2011).

Although it is usually impossible to inspect every record individually, summary statistics can provide a valuable overview of a dataset. The first step with numeric data is usually to produce a distribution histogram for each variable. Outliers—data points which are at the extremes of the distribution—can be identified and investigated individually. Some outliers may genuinely be extreme values, which should be included in later analysis, while others may be due to errors, and can legitimately be discarded. It is, of course, vital to decide upon a principled set of exclusion criteria. Other summary statistics such as variable means and ranges, and the production and inspection of scatterplots can also provide overviews of the data.

Other data manipulations which might need to be performed, depending upon the data and the analyses to be performed include, but are not limited to:

- *Scaling* of different types of data with very different ranges
- *Transformation* so that data has a normal distribution, for some statistical approaches
- *Identification of overlap* between different datasets
- *Identification of statistical correlation* between variables

Once the time-consuming data pre-processing step is complete, the researcher can turn to the more productive task of carrying out the analysis. The choice of algorithms to be used depends upon both the nature of the data and the question to be answered. It is often valuable to use several different algorithms, since individual algorithms have different strengths and weaknesses, and together can provide complementary information.

4 INFERENTIAL TECHNIQUES
4.1 Basic statistical analysis

Basic statistical approaches play two roles in the data mining process. As discussed above, the calculation of basic statistics such as the mean, median and range for each variable, provides a good initial understanding of the data, and helps to identify outliers and other unexpected values. In addition, where established statistical techniques are applicable to the analysis of a particular dataset, they should be tried before applying more heuristically based approaches. Statistical approaches are based on long-established and well-accepted principles, and provide measures of the probability that a particular observation would arise by chance. However, standard statistical approaches are not always applicable to large, noisy biological datasets. Many techniques assume that the data has a particular distribution—often a normal, or Gaussian distribution—and may fail if this distribution is not present. Some techniques have problems with missing or incomplete data. And sometimes it is simply not feasible to perform a statistical test on a very large dataset; heuristic approaches are sometimes necessary.

In this section we briefly discuss some of the most widely applicable basic statistical approaches to data mining, without going into technical details. There are many excellent textbooks on statistics for the biological sciences (Paulson, 2008).

In addition to assumptions about the distribution of the data, a major consideration when using statistical analysis is the type of data input and output by different techniques. Data may be:

- *Continuous*: real-valued numbers, such as the ratios produced by DNA microarrays;
- *Categorical*: each number represents a different category. For example, a protein may be coded as 0, nuclear; 1, cytoplasmic; 2, secreted. Categorical data may be recoded as a set of dichotomous variables (variables that take on one of only two possible values or "dummies") for input to some algorithms. The protein localisation data could be recoded as three variables, each of which takes the value 0 (no) or 1 (yes): nuclear, cytoplasmic and secreted;
- *Ordinal*: each number represents a category, but the categories have an inherent order. For example, data collection for a time-series analysis may be carried out on 0, Monday; 1, Tuesday; 2, Wednesday.

Different statistical techniques consume and produce different types of data (Table 2.1).

Table 2.1 Different Analytical Techniques are Suited to Different Types of Input Data and Provide Different Types of Output Data

Technique	Predictors	Output
Discriminant Analysis	Continuous or dummies	Categorical
Linear Regression, ANOVA	Continuous or dummies	Continuous
Logistic/multinomial regression	Continuous or dummies	Categorical
Ordinal Regression	Categorical or continuous	Ordinal
Time Series Analysis	Continuous or dummies	Continuous

4.2 Linear regression

A linear regression produces a "line of best fit" through a scatterplot. The criterion used to fit a line to a given dataset is the minimization of the sum of the squared deviation of each point about the line (the *residuals*). The output of the procedure is usually a visual image of the data and the line of best fit (Figure 2.3), a measure of the goodness of the fit of the line, R^2, ranging between 0.0 and 1.0, with 1.0 representing a perfect fit; and the equation of the line. The presence of outliers in the data can reduce the R^2 value. For a two-variable analysis the equation of a line is of the form $y = Ax + B$, and this equation can be used to predict the value of new observations.

Multiple regression analysis is a simple extension of linear regression to incorporate multiple variables. Multi-dimensional data is much harder to visualise than a simple two-dimensional scatterplot, but the application of multiple regression is exactly the same: to produce an equation that can be used to predict the value of new observations. This equation is of the form $y = B_1 * x_1 + B_2 * x_2 + B_3 * x_3 + \cdots + A$, where the Bs are the multipliers on the observations, x, and A is a constant.

Regression algorithms rely upon a number of assumptions: that the variables are on an interval scale (i.e. one unit change has the same meaning throughout the range of the data); that the residuals have a normal distribution; and that the residuals are independent of the predicted values. Most statistical software packages will allow the user to check whether these assumptions are met.

There are many different types of regression, including logistic, multinomial and ordinal variables. These algorithms take different types of input data, and fit different types of curves to the data (Figure 2.4).

4.3 Discriminant analysis

Discriminant analysis is a way to build classifiers: that is, the algorithm uses labelled training data to build a predictive model of group membership which can then be applied to new cases. While regression techniques produce a real value as output, discriminant analysis produces class labels. As with regression, discriminant analysis can be linear, attempting to find a straight line that separates the data into categories, or it can fit any of a variety of curves (Figure 2.5). It can be two dimensional or multidimensional; in higher

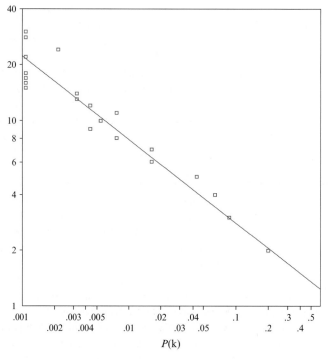

FIGURE 2.3

A linear regression draws a line of best fit through a scatterplot of the data. (For colour version of this figure, the reader is referred to the online version of this chapter.)

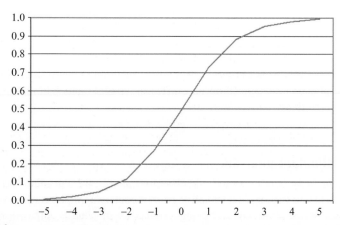

FIGURE 2.4

The logistic function is a smooth curve. (For colour version of this figure, the reader is referred to the online version of this chapter.)

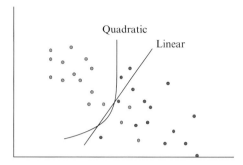

FIGURE 2.5

Discriminant analysis attempts to identify a boundary between groups in the data, which can then be used to classify new observations. The boundary may be linear or nonlinear; in this example both a linear and a quadratic line are fitted. (For colour version of this figure, the reader is referred to the online version of this chapter.)

dimensions the separating line becomes a plane, or more generally a *hyperplane*. Discriminant analysis also outputs an equation that can be used to classify new examples.

Discriminant analysis makes the assumptions that the variables are distributed normally, and that the within-group covariance matrices are equal. However, discriminant analysis is surprising robust to violation of these assumptions, and is usually a good first choice for classifier development.

> **Software Availability**
>
> R: http://www.r-project.org/. R is a statistical programming language. It has a fairly steep learning curve, but is extremely powerful. It has numerous libraries, including one for the analysis of biological data:
>
> Bioconductor: http://www.bioconductor.org/

4.4 Principal components analysis

The aim of much data mining is to identify the variables, representing real-world factors, which explain most of the variability in a dataset. Large datasets typically contain many variables describing each record, and the effects of variables may be nonlinear and interacting. To reduce the number of variables that must be considered, techniques for dimensionality reduction are often used.

Clustering is essentially a means of dimensionality reduction that involves retaining all of the data, but identifying commonalities that allow groups of data items to be treated together. An alternative approach is to discard some of the data, retaining only those features that contain the maximum information. Analysis of the data thus becomes less computationally demanding, and the results may be easier to understand and interpret.

The most oldest and most widely used of these is a statistical method called Principal Components Analysis (PCA). PCA takes an input matrix in which the rows are

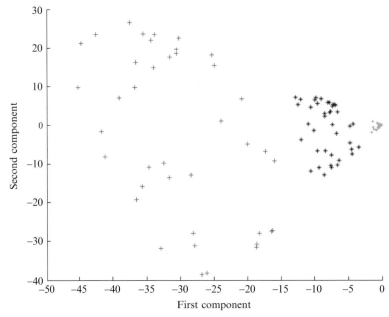

FIGURE 2.6

Output of a PCA. The first principal component is plotted against the second principal component, and different categories of case are indicated by colour. Image courtesy of Petter Strandmark, via Wikimedia Commons.[25] (For colour version of this figure, the reader is referred to the online version of this chapter.)

individual cases and the columns are the variables associated with the cases. These variables are usually inter-correlated. The aim of PCA is to extract the important information from the matrix and present it as a set of new orthogonal variables, the *principal components*, created by combining the existing variables. The principal components explain the variability in the data. Optimally, the first few principal components will explain most of the variability, and the rest of the principal components can be discarded. Further analysis can then be carried out using just a small number of principal components, instead of a large number of original variables. The mapping between the original matrix of variables and the principal components can be used to compute factor scores for new cases as they arise. The relationship between individual cases and the principal components can be visualised by displaying them as points on maps (Figure 2.6) (Abdi and Williams, 2010).

PCA has contributed to the analysis of large datasets in just about every aspect of microbiology. Some interesting recent applications include investigation of metagenomes in the human gut, both normal (Qin et al., 2010; Wu et al., 2011)

---

[25]http://commons.wikimedia.org/wiki/File:Kernel_pca_output.png.

and diseased (Schippa *et al*., 2010; Sobhani *et al*., 2011), the soil (Delmont *et al*., 2011); assessing the antibacterial activity of microbes (Wang *et al*., 2010); and many investigations into microbial community fingerprints (Illian *et al*., 2009; Xu *et al*., 2010; Zhang *et al*., 2010a).

5 MACHINE LEARNING TECHNIQUES

Statistical approaches have the advantages of being well established, with a solid mathematical foundation, and are generally well accepted by the scientific community. Wherever possible, statistical approaches should be tried on large, complex datasets. However, the price to pay for the strong foundation of statistical techniques is the limitations on the type of data that can be handled, and, often, the assumptions made about the distributions of the variables to be explored. When data are very large or messy, or there is no plausible hypothesis to be tested, heuristically based machine learning techniques can provide unique insights into the underlying biological processes.

5.1 Support vector machines

A support vector machine (SVM) is yet another type of classification algorithm (Boser *et al*., 1992). SVMs attempt to identify a separating hyperplane between classes, in a manner similar to discriminant analysis. It differs, however, from discriminant analysis in the way in which the hyperplane is selected. An SVM attempts to select the hyperplane that is in the middle of the gap between the categories, and therefore maximally far away from both classes of data. However, real data is rarely cleanly separable, and the SVM algorithm allows for some misclassifications. The proportion of misclassification is controlled by a user-adjustable parameter, known as the *soft margin*.

SVMs are characterised by the use of a *Kernal function* that adds an extra dimension to the data, essentially projecting it from a low-dimensional space into a higher-dimensional space. Data are more widely scattered in higher-dimensional spaces, and are therefore often more easily separable. It has been proven that for any data set there exists a Kernal function which will allow the data to be linearly separated (Noble, 2006), but the task of identifying this function is a black art, and kernals are usually chosen by trial and error. SVMs can be extended to handle more than two classes of data in a relatively straightforward manner (Lee *et al*., 2004b; Noble, 2004). SVMs also have the advantage of not assuming that the training data is normally distributed.

The advantages of SVMs for classifier construction mean that they have been widely used in a number of fields, including microbiology. They have been used for tasks such as predicting the subcellular localisation of proteins (Gardy *et al*., 2005; Rashid *et al*., 2007), gene finding (Krause *et al*., 2007), analysis of proteins (Rausch *et al*., 2005) and protein function classification (Cai *et al*., 2003).

SVMs are widely used in microarray analysis, exploring issues such as the nature of host-microbe interactions (Cummings and Relman, 2000), the prediction of mitochondrial proteins (Kumar *et al.*, 2006), prokaryotic gene finding (Krause *et al.*, 2007), protein functional classification (Cai *et al.*, 2003), protein subcellular localisation (Bhasin *et al.*, 2005; Gardy *et al.*, 2005; Gardy and Brinkman, 2006) and even the tracking of the source of microbes in heavily polluted water (Belanche-Muñoz and Blanch, 2008).

> **Software Availability**
>
> SVM^{light}: http://svmlight.joachims.org/. Free for scientific use; source code and binaries available.
>
> Gismo (Gene Identification Using a Support Vector Machine for ORF Classification): http://www.cebitec.uni-bielefeld.de/groups/brf/software/gismo/. Source code in Perl; requires local installation of Perl plus a number of Perl modules.
>
> SVMProt: http://jing.cz3.nus.edu.sg/cgi-bin/svmprot.cgi. Protein functional family prediction.

5.2 Hidden Markov models

Hidden Markov models (HMMs) were first introduced in the 1960s (Baum and Petrie, 1966), and have been applied to the analysis of time-dependent data in fields as such as cryptanalysis, speech recognition and speech synthesis. Their applicability to problems in bioinformatics became apparent in the late 1990s (Krogh, 1998). HMMs are frequently used for the statistical analysis of multiple DNA sequence alignments. They can be used to identify genomic features such as ORFs, insertions, deletions, substitutions and protein domains, amongst many others. HMMs can also be used to identify homologies; the widely used Pfam database (Punta *et al.*, 2012), for example, is a database of protein families identified using HMMs. HMMs can be significantly more accurate than the workhorse of sequence comparison tools, BLAST (Basic Local Alignment Search Tool), first produced in 1990 (Altschul *et al.*, 1990, 1997).

An HMM is a statistical model of a sequence. It consists of a library of symbols making up the sequence, and a set of *states* that an element of the sequence might occupy. Each state has a set of weighted *transition probabilities*: the probability of moving to a different state. A transition probability depends solely upon the previous state; states prior to the previous state have no effect on transition probabilities. An HMM also has a set of *emission probabilities*: the probability of producing a particular element of the sequence (Figure 2.7). A model is trained using known sequences to optimise the weights, and can then be applied to unknown sequences in order to make predictions. Since several paths through an HMM may produce the same sequence, paths are ranked by likelihood, by multiplying all of the probabilities together and taking the logarithm of the result. An algorithm known as the Viterbi algorithm (Forney, 1973) provides an optimal state sequence for many purposes.

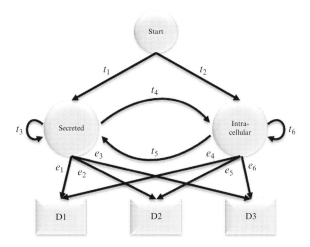

FIGURE 2.7

A very simple HMM to determine whether a protein is secreted or intracellular, based upon its sequence of domains, D1–D3. Variables t_1–t_6 are transmission probabilities; e_1–e_6 are emission probabilities.

There are several variants on the basic HMM, with slightly different functionality.

HMMs have recently been used for tasks such as detecting genomic islands, frequently associated with virulence, in newly sequenced bacterial genomes (Langille *et al.*, 2010); identifying and classifying secretome proteins (Craddock *et al.*, 2008), and promoter prediction in the *Chlamydia* genome (Malios *et al.*, 2009). HMMs have long been used for gene finding (Lukashin and Borodovsky, 1998; Azad and Borodovsky, 2004; Besemer and Borodovsky, 2005), including specifically identifying gene starts in microbial genomes (Besemer *et al.*, 2001).

A powerful approach to some problems is to combine HMMs with other bioinformatics tools. For example, (Snir and Tuller, 2009) combined an HMM with a Maximum Likelihood approach in order to analyse and model phylogenetic networks.

Software Availability

HMMER (http://hmmer.org/);

SAM (http://www.cse.ucsc.edu/research/compbio/sam.html).

PSI-BLAST (http://www.ncbi.nlm.nih.gov/Education/BLASTinfo/psi1.htm).

PFTOOLS (http://www.isrec.isb-sib.ch/profile/profile.html).

GeneMark: http://opal.biology.gatech.edu/GeneMark/.

5.3 Decision trees

Decision trees are visually similar to the graphical representation of HMMs, but operate on very different principles. A decision tree is a type of classifier, which takes a set of inputs describing individual data items, and classifies each item into one of a set of categories. Decision tree algorithms are trained using a set of input examples, each labelled with the category to which it belongs. The algorithms examine the input data to determine which variable best distinguishes between the categories, and which values of this variable are informative. This variable forms the root of the tree. The remaining variables are scrutinised for the next-most-informative variable, which generates the second level of the tree. The process continues until maximum separation between the output categories is achieved. Not all input variables will be included in the tree, so the decision trees also provide a means of feature selection.

For example, (Dieckmann and Malorny, 2011) used a decision tree to classify serovars of *Salmonella enterica* subsp. *enterica* using data from MALDI-TOF MS. Part of the resulting tree is redrawn below (Figure 2.8).

There are many decision tree algorithms, each of which uses different ways of identifying informative variables. One of the most widely used algorithms is C4.5 (Quinlan, 1993). C4.5 is freely available, although it is no longer supported. The successor of C4.5, C5.0 is only available as source code, which requires a compiler for the C language, and is therefore less accessible to the casual user.

The C4.5 algorithm uses a metric called entropy, which is derived from information theory (Shannon, 1948). Claude Shannon was an American mathematician and

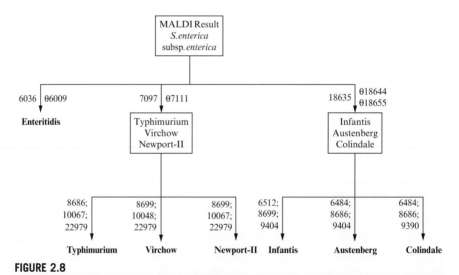

FIGURE 2.8

Decision tree for the classification of *Salmonella enterica* subsp. *enterica* serovars. The input data was generated using MALDI-TOF MS. Only part of the tree is shown.

Redrawn from Dieckmann and Malorny (2011).

electrical engineer who worked on communications theory. Amongst many other achievements, he produced a set of metrics that are widely used in many different fields. One of these metrics is entropy, which is essentially a measure of the randomness of a system (Figure 2.9). Entropy is the expected number of bits required to encode the classes, C_1 or C_2, of a randomly drawn member of a signal, S, under the optimal, shortest-length code:

$$\text{Entropy}(S) \equiv -p_{C1} \log_2 p_{C2} - p_{C1} \log_2 p_{C2},$$

where p_{C1} is the proportion of S having type C_1, and p_{C2} is the proportion of type C_2 (Figure 2.9). The information value of a variable, A, is calculated based upon its *information gain*: the expected reduction in entropy due to sorting on A.

$$\text{Gain}(S, A) \equiv \text{Entropy}(S) - \sum_{v \in \text{Values}(A)} \frac{|S_v|}{|S|} \text{Entropy}(S_v).$$

At each iteration of the algorithm, the information gain is calculated for each variable A in turn, and the variable which provides maximum information gain is selected as the best decision attribute for that node. For each value of A, a new descendant node is created, and the training examples are sorted to the nodes. If the training examples

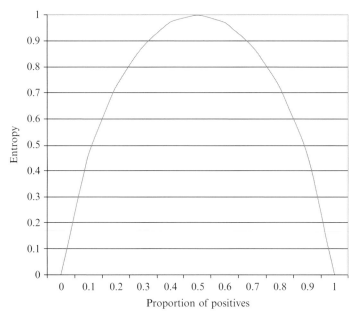

FIGURE 2.9

The entropy of a population increases with the proportion of positives, up to 50%, and then decreases smoothly. (For colour version of this figure, the reader is referred to the online version of this chapter.)

are perfectly sorted, the algorithm terminates; otherwise the same procedure is carried out for each of the new nodes.

The C4.5 algorithm has the advantage of producing a single classification for each data item, and because of its statistical basis, it is relatively robust to noisy data. It tends to produce short trees, with high information gain near the root, a generally desirable characteristic. Decision trees also carry out a form of feature selection, since only the most informative variables are included in the tree. From a practical point of view, the algorithm is easy to use, once the requisite data has been prepared, and it produces results that are easy to understand. Other widely used decision tree algorithms include the Chi-squared Automatic Interaction Detector (Kass, 1980), and Multivariate Adaptive Regression Splines (Friedman, 1991).

Decision trees are useful for data where the input variables are either continuous or categorical, and the outputs are categorical. They have the advantage of being able to classify data into any one of multiple categories, but they do require a relatively large amount of data. Decision trees can be turned into sets of rules, which can then be incorporated into computer programmes, allowing the automated application of a trained decision tree to new data as it is generated.

Decision trees have been widely used in microbiology research, in areas such as microbial identification (Rattray *et al.*, 1999; Ferdinand *et al.*, 2004; Dieckmann and Malorny, 2011), determination of the phylogenetic group of *Escherichia coli* (Clermont *et al.*, 2000), protein functional annotation (Azé *et al.*, 2007), classification of regulatory phenotype (Bachmann *et al.*, 2009), environmental monitoring and tracking the source of medically important microbes (Lyautey *et al.*, 2007, 2010; Ballesté *et al.*, 2010) and understanding transcriptional control (Singh *et al.*, 2005; Nannapaneni *et al.*, 2012).

> **Software Availability**
>
> C4.5: http://www.rulequest.com/Personal/.
>
> C5.0: http://rulequest.com/download.html (source code in C; will need to be compiled).
>
> Simple Decision Tree: http://sourceforge.net/projects/decisiontree/.

5.4 Clustering

Decision trees take a mass of data and try to sort it into discrete, meaningful categories. A similar approach is cluster analysis, which also attempts to group data into discrete categories, but without the aid of training data, and without necessarily specifying what these categories (clusters) mean.

The aim of clustering is to find subgroups within datasets that correspond to meaningful clusters *in vivo*. Cluster analysis is widely used in all fields of biology. Depending upon the dataset, clusters may have different interpretations. In a yeast, two-hybrid dataset (Fields and Song, 1989) a cluster may represent a protein complex (Krogan *et al.*, 2006), while in a more generalised interactome a cluster may represent any sort of functional relationship, such as a biological pathway (Hallinan et al., 2009). Decision trees are a supervised form of learning, whereas clustering is unsupervised. Supervised algorithms are trained using examples of known category, whereas unsupervised algorithms do not rely upon the existence of a labelled training set. There are, again, literally hundreds of clustering algorithms; a recent textbook runs to four volumes (Byrne and Uprichard, 2012), a degree of coverage which we shall not attempt to emulate here.

The idea of clustering is intuitively attractive. Biology is inherently modular, and when a researcher is confronted with a large mass of data, the ability to organise it into at least semi-coherent groups can simplify analysis tremendously. One of the major aims of much DNA microarray analysis, for example, is to identify groups of genes that have similar expression profiles either over time or over a range of conditions.

Just about any sort of data can be clustered, from gene expression measurements, to taxonomic relationships, to interaction networks. However, therein lies a problem. Many clustering algorithms, when asked to find n clusters in a dataset, will find n clusters, whether or not they actually exist. Furthermore, many clustering algorithms are stochastic, and may produce different cluster memberships when applied repeatedly to the same data, or when the same data is presented to the algorithm in a different order. As always with data mining, there is currently no substitute for human expertise when it comes to the evaluation of clustering results.

All clustering algorithms are based upon the concept of *distance* between records. Members of a cluster are presumed to be closer to each other, in some easily computationally interpreted way, than members of separate clusters. The most widely used distance metric in microbiology is undoubtedly generated by an all-against-all BLAST, which generates a matrix of similarity values between each pair of proteins in the input dataset.

Another very simple distance metric, applicable to items which are described in terms of strings of integers, is the Hamming distance (Hamming, 1950). Consider two proteins, represented in terms of the domains, D1–D9, which they contain. In this representation 0 indicates that the domain is not present, while 1 indicates that it is.

	D1	D2	D3	D4	D5	D6	D7	D8	D9
Protein 1	0	0	1	1	0	1	0	0	1
Protein 2	0	1	1	0	0	1	0	1	1

The proteins differ at three positions: D2, D4 and D8. The Hamming distance between them is therefore simply 3.

If the variables describing the item are real numbers, the Euclidean distance is the simplest distance metric. To calculate the Euclidean distance one subtracts the value of the variables at the same position in each vector, squares the result, sums all of the values, and takes the square root of the sum:

$$D_{A,B} = \sqrt{\sum_{i=1}^{n}(A_i - B_i)^2},$$

where A and B are two vectors of variables, indexed by i, and n variables long. Squaring the result of the variable subtraction and then taking the square root of the final sum means that only the absolute values of the differences are considered.

Consider two proteins for which a number of features have been measured:

	Feature 1	Feature 2	Feature 3	Feature 4
Protein 1	20.8	153.2	23.5	54.8
Protein 2	25.4	120.0	22.8	65.2

The distance, D, between these proteins, using this representation, is:

$$D = \sqrt{(20.8 - 24.4)^2 + (153.2 - 120.0)^2 + (23.5 - 22.8)^2 + (54.8 - 65.2)^2}$$
$$= 10.4.$$

There are, of course, many other possible distance metrics that can be applied to microbiological data. New metrics are being developed constantly, as new types of data are generated and old types re-evaluated.

There are a number of clustering algorithms that are widely used because they are easy to understand and interpret. One of the most popular is the k-nearest neighbour (k-nn) algorithm (Fukunaga and Narendra, 1975). As with most clustering algorithms, k-nn starts with a matrix of distances between every item in the dataset. The dataset must contain some items whose classification is known. The single parameter k is chosen to be an odd number, and each item in the dataset is simply assigned to the category held by the majority of its k nearest neighbours; that is, the k data items which are closest to it using the chosen metric.

A slightly more sophisticated algorithm is k-means (Hartigan and Wong, 1979). For this algorithm, the user chooses the number of clusters, k, believed to exist in the data. The mean, or *centroid* (centre of mass), of each cluster is assigned, either on the basis of prior knowledge, or by selecting well-spaced items randomly from the data to be clustered. The clustering process is iterative: the input data are read one at a time, and each item is assigned to the cluster to whose centroid it is closest. The centroids of the clusters are then re-calculated. This process continues until the centroids no longer change. This algorithm is sensitive to both the number of clusters used, and to the initial centroid assignment, so it is generally wise to repeat it multiple times, varying both of these factors and examining the resulting clusters for plausibility. Fortunately, this simple algorithm is very fast, even with large datasets, so the process of automating a large number of runs is straightforward and efficient.

Once cluster membership is determined, it can be used for a number of purposes. One of the most common is the prediction of the characteristics of unknown proteins based upon their cluster membership: the so-called guilt-by-association approach (Altschuler *et al.*, 2000). Cluster membership has been widely used in this way to predict protein function (Sharan *et al.*, 2007; Mostafavi *et al.*, 2008).

Cluster analysis has been applied within microbiology to address a wide range of questions. Perhaps the most common application is to the analysis of the composition of microbial communities under natural conditions (Noble *et al.*, 1997; Juck *et al.*, 2000; Zhang and Fang, 2000; Blackwood *et al.*, 2003), when disturbed by agriculture or pollution (Rooney-Varga *et al.*, 1999; Juck *et al.*, 2000), and in disease (Frank *et al.*, 2007). Cluster analysis has also been used for the identification of transcriptional modules (Leyfer, 2005); investigating the evolution of pathogenicity (Keim *et al.*, 2000; Tettelin *et al.*, 2005); and gene identification and protein classification (Yooseph *et al.*, 2008).

Cluster analysis algorithms are not always designed to produce a neat set of clusters with clearly defined membership. Hierarchical clustering methods, of which there are many, generate a tree, or dendrogram, in which different levels of the tree represent different granularities of the clustering. Data items can be assembled into a tree by *agglomeration*, in which the items that are closest together, on the basis of the distance metric chosen, are iteratively grouped together. Alternately, a divisive procedure can be used, with the entire dataset initially considered as a single cluster, and then divided into successively smaller clusters on the basis of distance between the cluster members or cluster centroids.

A hierarchical tree terminates in a set of leaf nodes, each of which contains a single member of the original dataset, making it valuable for examining the relationships between individuals. However, a cluster tree can also be thresholded at higher levels of granularity, to investigate relationships between groups of individuals (Figure 2.10). In many phylogenetic trees (although not in Figure 2.10) the length of the vertical lines represents evolutionary time since the last devolutionary split.

Probably the most familiar application of hierarchical clustering in microbiology is for the construction of phylogenetic trees, which, we hope, reflect the evolutionary relationships between organisms. Phylogenetic trees are usually built on the basis of distances calculated between hypervariable regions of the genome. The most widely used regions are the 16S rRNA genes, which have been used since the late 1970s (Woese and Fox, 1977).

Although the genes used for phylogenetic analysis have not changed for 30 years, the technologies used to obtain genetic data are constantly changing and developing, necessitating the development of new approaches to analysis. Most recently, the advent of Next Generation sequencing has made possible the generation of huge amounts of sequence data, quickly and cheaply, albeit in the form of short reads of 100–200 bp. Such data requires new techniques for cluster analysis, which take into account the nature of the data being analysed (Huse *et al.*, 2010; Lemos *et al.*, 2011; Foster *et al.*, 2012).

Another very widely used application of cluster analysis is the investigation of time-course DNA microarray data (Figure 2.11). The aim of many microarray

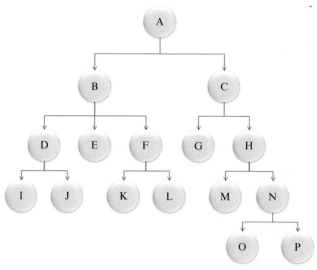

FIGURE 2.10

An example of a dendrogram, such as may be output by a hierarchical clustering algorithm.

experiments is to explore patterns of gene expression over time, or over a range of different experimental conditions. Genes with similar patterns of expression may be regulated by the same transcription factor (TF), or may participate in the same biological process. Clustering of microarray data is therefore a useful mechanism of hypothesis generation.

There are, unsurprisingly, hundreds of clustering algorithms that have been applied to microarray data. Cluster analysis is simple, easy to understand, and potentially very powerful. There is a clustering algorithm for almost any type of data, and the clustering process can help to uncover biologically relevant patterns of relationships in otherwise unmanageably large datasets. However, there are a number of caveats that must be borne in mind when choosing and applying a clustering algorithm.

- *Number of clusters:* Many clustering algorithms require the user to specify how many clusters exist in the data, and will return that number of clusters, whether or not the result is biologically plausible. The k-means is one such algorithm. Unless there is other evidence for the existence of a specific number of clusters in the data the results of such an algorithm should be critically evaluated.
- *Stochasticity:* Clustering algorithms often incorporate an element of chance, with regards to issues such as which data item is selected in any iteration of the algorithm. Stochastic algorithms may produce different cluster memberships in different runs. Such algorithms should therefore be run repeatedly and the results combined.
- *Sensitivity to ordering:* A similar issue is that some algorithms can predict different cluster memberships depending upon the ordering of the input data.

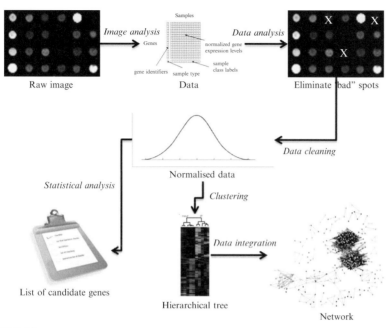

FIGURE 2.11

A simplistic view of the microarray analysis pipeline. (For colour version of this figure, the reader is referred to the online version of this chapter.)

Modified from Boutros and Okey (2005).

- *Visual representation:* The way in which clustering results are visualised can affect the way in which a user interprets them. This is particularly true with dendrograms, in which the layout of the tree may imply a spurious closeness or similarity between nodes. For example, in Figure 2.5 the layout of the tree may be interpreted as meaning that node F is closer to node G than is node D. However, in this type of dendrogram each decision point should be interpreted as essentially a swivel; nodes D, E and F are all equally distant from node G.

In view of the issues discussed above, clustering algorithms should be selected with an eye to their strengths and weaknesses. Stochastic algorithms should be run repeatedly, and the results of all cluster analyses should be examined carefully. This is not to imply that cluster analysis cannot be a very valuable data mining tool; but simply that it should be used intelligently.

Software Availability
ScanAlyze, Cluster and Treeview: http://rana.lbl.gov/EisenSoftware.htm.

5.5 Artificial neural networks

Artificial neural networks (ANNs) are a machine learning approach which dates back to the 1940s (McCulloch and Pitts, 1943), although they did not become widely useful until the 1980s (Hecht-Nielsen, 1989). Neural networks are inspired by the human brain, and are composed of artificial "neurons", which are connected by weighted edges. There are far too many different neural network algorithms to address in detail here, and there are many excellent textbooks on neural networks (Bishop, 2007; Haykin, 2008). Here we restrict ourselves to what is undoubtedly the most widely used ANN algorithm, the multi-layer perceptron (MLP). In an MLP, the neurons are arranged in layers, with input nodes which take the input data, output neurons which output the results of the computation, and so-called hidden neurons, which are neither input not output (Figure 2.12). An MLP may have one or more layers of hidden neurons.

Each neuron calculates the sum of all of its inputs, runs this sum through a squashing function, and outputs the result. Learning in the ANN consists of iteratively modifying the weights on the edges in response to errors in classifying labelled training data, a process which is usually carried out using the *backpropagation* algorithm (Hecht-Nielsen, 1989).

A very common squashing function for the neurons is the *sigmoid* function, which is described by the equation $y = 1/(1 + e^{-x})$ (Figure 2.13).

A function such as the sigmoid can takes input values of any magnitude, and output a value of between 0 and 1, thereby keeping the results of the computation performed by the ANN within prescribed limits. In the case of the example network in Figure 2.12, for identifying protein location, the output of the single output node would be between 0 and 1, and would usually be thresholded at 0.5. If the training

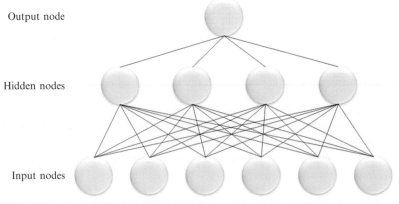

FIGURE 2.12

A fully connected feed-forward multi-layer perceptron. This example has six input nodes, four hidden nodes and one output node, but the architecture is flexible based upon the needs of the problem.

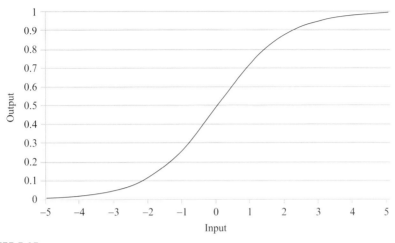

FIGURE 2.13

The sigmoid function.

data were labelled with, for example, 0 for proteins with a cytoplasmic location and 1 for secreted proteins, and set of variables which led to an output less than 0.5 would be classified as having a cytoplasmic location, while any with an output of more than 0.5 would be classified as secreted. It can be demonstrated that a neural network can learn any function with arbitrary precision but, as so often the case with machine learning algorithms, the selection of requisite architecture and training regime is a "black art", usually addressed using trial and error.

Neural networks can be a very powerful approach to classification and modelling of complex systems. They do require a lot of data; one rule of thumb is that there should be at least five cases in the training set for each weight in the network. Dataset size is, however, often a moot point with microbiological data. Input data can be real valued, categorical or binary, making ANNs a flexible approach to many problems. One criticism often directed at ANNs is that they are "black boxes". Unlike classifiers such as decision trees, it is not easy (although not impossible) to untangle the relative importance of the variables that contribute to the classification.

Neural networks have been used in microbiology to address a very wide range of problems, such as the estimation of hydrogen production by genetically modified *E. coli* (Rosales-Colunga et al., 2010); optimization of biomolecule production (Singh et al., 2009; Nelofer et al., 2012); predicting bacterial community assemblages (Larsen et al., 2012); elucidating the relationship between growth and environmental factors in *Staphylococcus aureus* (Fernández-Navarro et al., 2010); optimising fermentation conditions for *E. coli* (Silva et al., 2012) and predicting essential genes in microbial genomes (Palaniappan and Mukherjee, 2011) (this research also incorporated SVMs and decision trees). There have even been attempts to build ANNs using microbes (Ozasa et al., 2009).

> **Software Availability**
> JavaNNS: http://www.ra.cs.uni-tuebingen.de/software/JavaNNS/.

5.6 Ontologies and text mining

Publically available databases are an unparalleled resource for the microbiology community. They have, however, a number of problems. As already mentioned, much of the data in these databases are generated using high-throughput technologies, and significant amounts of the data have not been subjected to human curation. The data are noisy and incomplete, and often the proportions of false positive and true positive identifications and interactions are unknown. Further, the data are presented without context; we may know that a protein–protein interaction has been identified using a yeast two-hybrid approach, but we generally do not know what, if any, hypothesis the experiment which generated the data was designed to investigate or, perhaps most importantly, how the originating researcher interpreted the results. The most reliable data comes, indisputably, from the peer-reviewed literature.

Unfortunately, keeping up with the literature is impossible for a working scientist. According to an editorial in *Nature*, compiled from user input acquired using the social networking tool Twitter,[4] we have sequenced approximately $1 \times 10^{-22}\%$ of the DNA on earth: "the fraction of microbial diversity that we have sampled to date is effectively zero" (Microbiology by Numbers [Editorial], 2011). Even so, thousands of articles are published every month, and the rate of publication is increasing exponentially (Hunter and Cohen, 2006).

The title "last man to know everything" has been variously applied to a number of people, including Thomas Young (1773–1829) (Robinson, 2007), Athanasius Kircher (1601 or 1602–1680) (Findlen, 2004), Joseph Leidy (1823–1891) (Warren, 1998) and, of course, Gottfried Leibniz (1646–1716) (Fentress, 1914), all of whom worked before the middle of the 19th century. "Knowing everything" is no longer possible. Even in specialised sub-fields it is not feasible to scan all of the relevant journals, identify papers which may be relevant to current or future work, extract and understand the important findings, and organise the results in a way which can be easily accessed at need. The concept of automated text mining is therefore extremely attractive, and the application of text mining to genomics has been an active area of research for at least 20 years (Zweigenbaum *et al.*, 2007).

An important concept in literature mining, as in many other aspects of bioinformatics (and, indeed, many other fields) is that of an *ontology*. An ontology is a working conceptual model of the entities which exist in a given domain, and their interactions (Gruber, 1993; Stevens *et al.*, 2000). The basis of any ontology is a *structured vocabulary*: a list of terms that must be used to describe the entities in a domain.

---

[4] https://twitter.com/.

For example, in molecular biology the term "gene" is particularly vague. It may be defined as: an ORF; a protein coding sequence (CDS); a unit of heredity; and so on. In fact, in a large proportion of the technical literature, the term "gene" is used without definition, leaving the interpretation up to the reader. If that reader is a human, this approach is generally workable; a human will know whether the paper under consideration deals with, for example, prokaryotes or eukaryotes, and will understand the implications of this distinction for the application of the term. The fact that prokaryotic genes frequently overlap, do not contain introns, may occur in operons, and are not associated in linear chromosomes is part of the inferred knowledge of the reader, and need not be stated. However, if the "reader" is a computational algorithm, none of this knowledge can be assumed.

Structured vocabularies are a first step towards making text interpretable to computers. A structured vocabulary may mandate, for example, that the term "CDS" must always be used to mean "a sequence of nucleotides which codes for an mRNA which codes for a protein, or part thereof". As anyone who has ever used a Web form with drop-down boxes knows, a structured vocabulary prevents confusion due to the use of different terminology, or even the mis-typing of an agreed terminology.

Ontologies are far more, however, than just structured vocabularies. A structured vocabulary ensures consistent naming of entities in a domain, but entities do not exist in splendid isolation. Interactions between entities are the core of any complex system. In an ontology, entities and the interactions between them are annotated, again using a standard terminology (Figure 2.14). The presence of these annotations, with their strictly defined meanings, means that computational algorithms can reason over a dataset represented as a graph, constructed using an ontology, extracting relationships and generating hypotheses which were not previously apparent. This approach is particularly valuable for very large graphs, which are hard to display on a computer screen, and even harder for a human to comprehend once they are displayed. Given the necessarily complex and redundant language used in most biology papers, ontologies are clearly valuable for literature mining in general, and for biology in particular (Jensen and Bork, 2010).

Because of the value of ontologies to the representation and analysis of large datasets, there are multiple community-based standards organisations which aim to define agreed standards for various domains of biology. The umbrella organisation for this effort is the OBO (Open Biological and Biomedical Ontologies) foundry[5] (Smith *et al.*, 2007). This organisation describes itself as "a collaborative experiment involving developers of science-based ontologies who are establishing a set of principles for ontology development with the goal of creating a suite of orthogonal interoperable reference ontologies in the biomedical domain". Any researcher is welcome to participate in ontology development for their specific area, and at the time of writing there are eight active domain-specific ontologies and 94 candidate ontologies, ranging from broad domains such as "cell type" to highly specific areas such as "*Dictyostelium discoides* anatomy". Perhaps the most widely known ontology amongst

---

[5] http://obofoundry.org/.

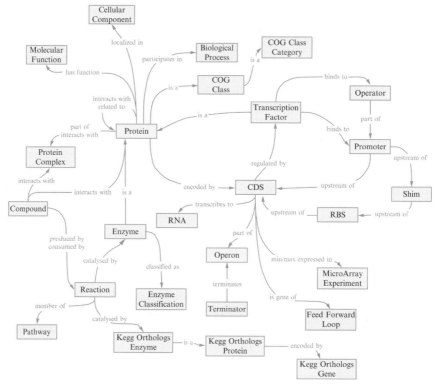

FIGURE 2.14

Graphical depiction of the ontology underlying BacillOndex, a semantically annotated *Bacillus subtilis*-specific knowledge base constructed using the Ondex data integration system.

working microbiologists is the GO[6] (Ashburner *et al.*, 2000), which provides a controlled vocabulary for the *biological process*, *cellular component* and *molecular function* of genes from a wide range of species.

In a biological context, text mining has perhaps been most widely applied in the field of systems biology, whose practitioners endeavour to understand biological systems in context, by integrating *in vivo* experiments with computational modelling and theory in an iterative manner.[7] In order to achieve this aim, the integration of large amounts of data is necessary, particularly when constructing simulateable models of biological systems, which generally require extensive parameterisation. Biologically reasonable values for model parameters are hard to find (Gutenkunst *et al.*, 2007), and must often be extracted from published literature, rather than from databases. As an alternative to dedicating weeks of valuable postgraduate student

---

[6]http://www.geneontology.org/.
[7]Definition adapted from http://www.bbsrc.ac.uk/web/FILES/Publications/systems_biology.pdf.

time to the sifting of journals, the prospect of automated literature mining is alluring, even if the process only generates a list of candidate data which must be further refined manually.

For the purpose of computational feasibility, text mining is often performed only on the abstracts of journal articles, although the continued increase in computational power, and particularly the availability of eScience approaches such as Cloud computing (discussed in more detail below) is rapidly making it more feasible to search large corpuses of entire articles. Abstracts also have the advantage of usually being freely available and relatively easy to download, making the generation of text mining datasets relatively straightforward. Moreover, running an exhaustive analysis once, and storing the results in a database can also reduce the demand for compute resources for text mining. The text mining algorithms need then only be re-run on new articles as they become available. This approach has been applied to the creation of knowledge bases on topics such as the identification of bacterial enteropathogens (Zaremba *et al.*, 2009), surveillance of the literature in the service of infectious disease control (Sintchenko *et al.*, 2009), toxin–antitoxin loci in bacteria and archaea (Shao *et al.*, 2011), information about integrative and conjugative elements found in bacteria (Bi *et al.*, 2012), and many others.

The range of algorithms applied to text mining in biology, and the details of their use, are more than extensive enough to warrant a review of their own, and there have been several useful contributions (Zweigenbaum *et al.*, 2007; Evans and Rzhetsky, 2011; Ceci *et al.*, 2012).

The simplest approach to text mining is the key word search. Although this method is trivially easy to implement, it generally performs poorly on molecular biology articles, due to the complexity and redundancy in the terminology used in the biomedical literature. The same concepts may be referred to in multiple ways, while multiple concepts may have the same name. This problem is particularly acute when it comes to gene identifiers; every database has its own form of identifier, and a paper may use any of the many "standards". Gene names may change over time, and older papers often use different aliases from newer ones.

One approach to dealing with the retrieval of information from sources that may use different terminology for the same concepts is the use of MeSH (Medical Subject Headings).[8] MeSH is a controlled vocabulary, used for indexing articles in PubMed. The terms are organised hierarchically, so that a user can search at different levels of specificity (Figure 2.15).

More complex approaches use semantics to try to parse papers into linguistically meaningful units. The most widely used approach to literature mining is Natural Language Processing (NLP), which has been an active area of research for several decades (Manning and Schutze, 1999). NLP has three parts: information retrieval, assignment of semantics and information extraction.

---

[8] http://www.ncbi.nlm.nih.gov/mesh/.

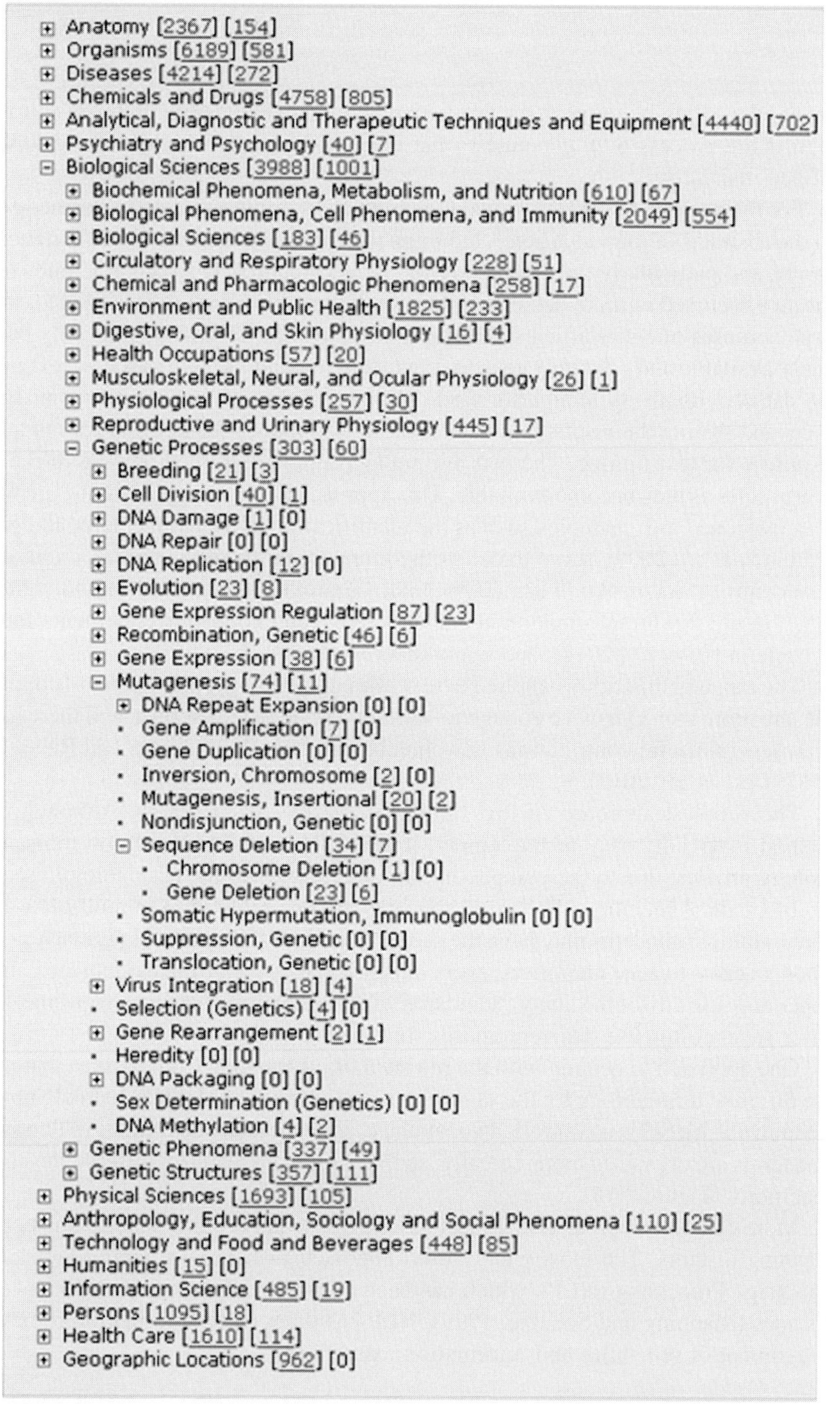

- Anatomy [2367] [154]
- Organisms [6189] [581]
- Diseases [4214] [272]
- Chemicals and Drugs [4758] [805]
- Analytical, Diagnostic and Therapeutic Techniques and Equipment [4440] [702]
- Psychiatry and Psychology [40] [7]
- Biological Sciences [3988] [1001]
  - Biochemical Phenomena, Metabolism, and Nutrition [610] [67]
  - Biological Phenomena, Cell Phenomena, and Immunity [2049] [554]
  - Biological Sciences [183] [46]
  - Circulatory and Respiratory Physiology [228] [51]
  - Chemical and Pharmacologic Phenomena [258] [17]
  - Environment and Public Health [1825] [233]
  - Digestive, Oral, and Skin Physiology [16] [4]
  - Health Occupations [57] [20]
  - Musculoskeletal, Neural, and Ocular Physiology [26] [1]
  - Physiological Processes [257] [30]
  - Reproductive and Urinary Physiology [445] [17]
  - Genetic Processes [303] [60]
    - Breeding [21] [3]
    - Cell Division [40] [1]
    - DNA Damage [1] [0]
    - DNA Repair [0] [0]
    - DNA Replication [12] [0]
    - Evolution [23] [8]
    - Gene Expression Regulation [87] [23]
    - Recombination, Genetic [46] [6]
    - Gene Expression [38] [6]
    - Mutagenesis [74] [11]
      - DNA Repeat Expansion [0] [0]
      - Gene Amplification [7] [0]
      - Gene Duplication [0] [0]
      - Inversion, Chromosome [2] [0]
      - Mutagenesis, Insertional [20] [2]
      - Nondisjunction, Genetic [0] [0]
      - Sequence Deletion [34] [7]
        - Chromosome Deletion [1] [0]
        - Gene Deletion [23] [6]
      - Somatic Hypermutation, Immunoglobulin [0] [0]
      - Suppression, Genetic [0] [0]
      - Translocation, Genetic [0] [0]
    - Virus Integration [18] [4]
    - Selection (Genetics) [4] [0]
    - Gene Rearrangement [2] [1]
    - Heredity [0] [0]
    - DNA Packaging [0] [0]
    - Sex Determination (Genetics) [0] [0]
    - DNA Methylation [4] [2]
  - Genetic Phenomena [337] [49]
  - Genetic Structures [357] [111]
- Physical Sciences [1693] [105]
- Anthropology, Education, Sociology and Social Phenomena [110] [25]
- Technology and Food and Beverages [448] [85]
- Humanities [15] [0]
- Information Science [485] [19]
- Persons [1095] [18]
- Health Care [1610] [114]
- Geographic Locations [962] [0]

FIGURE 2.15

MeSH tree browser view of the *Brucella* literature. (For colour version of this figure, the reader is referred to the online version of this chapter.)

From Xiang et al. *(2006).*

Information retrieval is simply the acquisition of documents from repositories. Anyone who has used PubMed, or a Web-based search engine has used standard information retrieval methods. Semantics—the assignment of meaning—is a non-trivial exercise, because of the aforementioned complexity of biological terminology. Assigning semantics is usually done using ontologies. Once a document is processed and indexed, information can be extracted by means of a database query, or some more specialised algorithm. An excellent review of the intersection between genomics and natural language processing can be found in (Yandell and Majoro, 2002).

One widely used, ontology-based tool is TextPresso, from the Generic Software Components for Model Organism Databases.[9] TextPresso splits papers into sentences, and then marks words or phrases with XML tags, derived from a specifically developed ontology. These semantically tagged snippets are stored in a database, which can be searched by keyword or category. The use of a tool such as TextPresso, however, requires significantly more technical ability than does a keyword search. As always, there is a trade-off between simplicity and power.

In order to address this problem, Web-based applications incorporating a range of algorithms are becoming increasingly popular; a recent review of 28 such tools has led to the construction of an overview site, based at the NCBI, and dedicated to tracking existing systems and future advances in the field of biomedical literature search (Lu, 2011).

Text mining has been applied to microbiology research to address just about every conceivable question, either alone or in combination with other data mining approaches. Some interesting recent examples include: automated inference of microorganism habitat (Kolluru *et al.*, 2011); exploring the dynamics of relationships between pathogens and infectious diseases (Sintchenko *et al.*, 2010); identifying viruses and bacteria with the potential to be used as bioterrorism weapons (Hu *et al.*, 2008); and the identification of molecules with potential pharmacological action (Sarker *et al.*, 2012).

> **Software Availability**
>
> PATRIC (Pathosystems Resource Integration Center): http://patricbrc.vbi.vt.edu/portal/portal/patric/Home (Includes a knowledge base constructed using text mining, plus several other valuable tools).
>
> Anni2.1: http://biosemantics.org/index.php?page=anni-2-0.
>
> NCBI list of biomedical text mining Web sites: http://www.ncbi.nlm.nih.gov/CBBresearch/Lu/search/.
>
> TextPresso: http://www.gmod.org/wiki/Textpresso.

---

[9]http://www.gmod.org/wiki/Main_Page.

5.7 Data integration and network analysis

High-throughput biological datasets tend to be large, noisy and incomplete. Different types of data are stored in different databases, and may have very different file formats. Different databases may also use different identifiers for the same protein, while gene and protein names and aliases may overlap. All of these problems mean that accessing all of the available data about an organism, protein or process of interest is a time-consuming, tedious and error-prone business if performed manually.

A widely used approach to the analysis of large amounts of biological data is data integration: bringing together data from a range of sources to produce a network, in which the nodes usually represent genes or gene products, and the edges between nodes indicate some type of interaction (Yandell and Majoro, 2002; Lee et al., 2004a; Hallinan and Wipat, 2006). The meaning of an edge depends upon the type of data being analysed; if the underlying dataset contains, for example, physical protein–protein binding data, such as that derived from yeast two-hybrid analysis, an edge between two nodes denotes that those proteins physically bind to each other, at least under experimental conditions. If the data, in contrast, are TF binding sites, edges represent transcriptional control.

Networks have the advantage of being easily interpretable. The network representation is familiar to most people, and the concept of nodes representing individuals and edges representing interactions is intuitively obvious. From a more technical point of view, networks can bring together large amounts of disparate data and present it in context, in a format which is easily browsable, or which can be analysed computationally. For example, the SinI/R operon in *Bacillus subtilis* lies at the heart of the organism's cell fate decision, and as such has been widely studied. Data about the relationships between these genes are scattered amongst multiple databases, and hundreds of papers. Using data integration, these data can be combined to produce a single picture of the current state of knowledge about this important system (Figure 2.16).

Networks can be constructed using data from a single type of experiment (e.g. genetic relationship inferred from synthetic lethal mutation data) or may be composed of data derived from many sources. In the former case an edge has a clear meaning: the two proteins which it joins are synthetic lethals. In the latter an edge represents any functional relationship between the two proteins, possibly based upon several different types of interaction.

Because of the noisiness of high-throughput data, it is often desirable to have an indication of the reliability of the data; that is, the probability that an edge present in a computationally integrated network actually exists *in vivo*. This probability is usually estimated by comparing the set of interactions present in the integrated network with those in a high-confidence, usually manually curated *Gold Standard* network. Commonly used Gold Standard datasets include KEGG,[10] the MIPS database[11] and the GO[12] (Lee et al., 2004a).

---

[10]http://www.genome.jp/kegg/.
[11]http://www.helmholtz-muenchen.de/en/ibis.
[12]http://www.geneontology.org/.

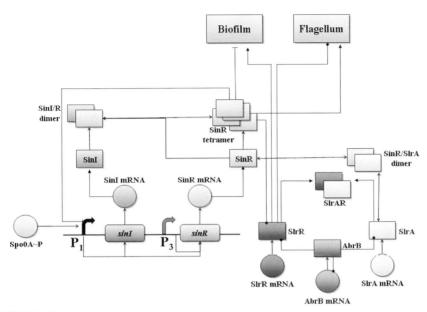

FIGURE 2.16

Network representation of the SinI/R system in Bacillus subtilis, indicating important regulatory relationships. (For colour version of this figure, the reader is referred to the online version of this chapter.)

There are several problems with the use of a Gold Standard for edge confidence estimation. Most importantly, all potential Gold Standards, no matter how well curated, are incomplete, and may contain errors (Cusick *et al.*, 2008). After all, if the Gold Standard was complete we would have no need to build integrated networks in the first place. Different Gold Standards produce different weightings, because they contain different data. Recent research has investigated the use of problem relevance in weighting edges (James *et al.*, 2009) and the assessment of edge probability without the use of a Gold Standard (Weile *et al.*, 2012), but such approaches are still not widely used.

There are many possible sources of data for the construction of integrated interaction networks (*interactomes*). Up-to-date information about publically available databases is available in the annual Nucleic Acids Research Database issue, published in January of each year. The 2012 issue details 1380 databases (Galperin and Fernández-Suárez, 2012). Not all of these databases contain data relevant to microbes, and those that do generally cover a limited range of species. Although there is a relatively large amount of data for model species such as *S. cerevisiae*, the amount of high-throughput data available for less well-studied species may be limited. Several of the major databases include data from multiple single-data-type sources. BioGrid[13] (Stark *et al.*, 2011), for example, is widely used, incorporates data

---

[13] http://thebiogrid.org/.

from a range of other databases, and is manually curated. Data in BioGrid is labelled with the PubMed ID of the publication from which it is derived, and is tagged with annotations, drawn from a controlled vocabulary, indicating the type of experiment from which it was generated (e.g. "Affinity Capture-mRNA", "Two-hybrid" and "Co-localization"). Another useful database is String[14] (Szklarczyk et al., 2011), which contains data about protein–protein interactions derived from a variety of microbial species. Some of the data in STRING are computationally generated, and may not be as reliable as human-curated data. The Microbial Protein Interaction database[15] (Goll et al., 2008) aims to collect and provide all known physical microbial interactions.

Despite the relative dearth of data for some microbial species, interactome analysis can be valuable even for data generated completely in-house. In particular, the ability to combine microarray data with other information, such as that about KEGG pathways or GO annotations can provide an entirely new perspective on the functional principles and dynamics of an entire cellular system (Tseng et al., 2012). An interesting recent review on the integration of multiple microbial "omics" datasets is provided by Zhang and colleagues (Zhang et al., 2010b), while Hallinan and co-workers discuss both microbial network integration and analysis (Hallinan et al., 2011).

Interactome analysis has been widely used in microbiology for predicting the function of un-annotated proteins. An integrated network is constructed, and then subjected to cluster analysis. A cluster, in a network context, can be defined as "a group of nodes which are more tightly connected to each other than to the rest of the network" (Hallinan et al., 2009). Most of the algorithms used for clustering non-network data can be adapted for clustering networks. Following clustering, the biological function of unknown proteins can be predicted using a "guilt-by-association" approach; proteins which occur in a cluster dominated by proteins with a single, known function are deemed to be likely to also have that function.

There are many algorithms for using interactomes for inferring the function of co-clustered genes (Wang and Marcotte, 2010). These are described in more detail in the Case Study (below).

Integrated networks can be used for a variety of other tasks, such as generating hypotheses about gene function, analysing gene lists and prioritising lists of genes for further functional assays. One such application, which has been the subject of considerable interest over the last decade or so, is the identification of network *motifs*. Network motifs are sets of small numbers of nodes, usually three to five, connected in a particular manner. They are assumed to perform a specific function, such as the amplification or damping of a specific signal, via a feed-forward or feedback loop, such as that represented by the *lac* operon in *E. coli*.

Motifs are believed to afford a mapping between network topology and real biological dynamics; the hope is that the time course behaviour of a large, complex

---

[14] http://string-db.org/.
[15] http://jcvi.org/mpidb/about.php.

network can be analysed, at least in part, by breaking it down into motifs whose behaviour is simple and well understood.

A number of motifs have been demonstrated to be statistically over-represented in metabolic and transcriptional networks. Motifs such as small feed-forward loops, single input modules and cycles have been identified as over-represented compared with equivalent randomly connected networks in *E. coli* (Shen-Orr *et al.*, 2002; Dobrin *et al.*, 2004) and *S. cerevisiae* (Wuchty *et al.*, 2003). It is often assumed that the motifs are over-represented because of positive selection pressure. However, Mazurie *et al.* (2005) compared over-represented motifs in the transcriptional network of *S. cerevisiae* with those in a number of other hemiascomycetes, and concluded that the regulatory processes for the biological function under consideration were dependent upon post-transcriptional regulatory mechanisms rather than transcriptional regulation by network motifs. They concluded that the presence of motifs is unlikely to provide a selective advantage to the organism, possibly because they are deeply embedded in the rest of a complex network of genetic interactions. Despite this controversy, the interactomes of a wide range of microbial genomes have been searched for network motifs in an attempt to understand the genome-wide, systems-level dynamics of the functional networks (Herrgård, 2004; Stelling, 2004; Gelfand, 2006; Janga and Collado-Vides, 2007; Ravcheev *et al.*, 2011)

Software Availability

Cytoscape: http://www.cytoscape.org/.

Ondex: http://www.ondex.org/.

STRING: http://string-db.org/.

GeneMania: http://www.genemania.org (Yeast only).

BioPixie: http://pixie.princeton.edu/pixie/ (Yeast only).

---

6 THE ROLE OF eSCIENCE

One of the most promising approaches to tackling very large data sets is that of global, collaborative analysis. Exchanging data and results across social, political and technological boundaries is a challenging process, both socially and technologically. Over the last decade there has been increasing research into, and use of, eScience. The term eScience refers to collaborative, global science that is performed *in silico* with a computational infrastructure (Luciano and Stevens, 2007). The major umbrella technologies supporting eScience are Grid and Cloud computing.

Grid computing was named by analogy to the electricity grid. The aim of Grid computing is that plugging in to worldwide compute resources should be as easy as accessing electricity (Kesselman and Foster, 1998). Grids are characterised by their geographic distribution: datasets and compute resources are held in different geographic locations, are usually maintained locally, and are connected via high-throughput networking. The elements of a Grid are generally heterogeneous, adding compatibility issues to those of data transfer, integration and maintenance. A Grid architecture requires appropriate protocols, services, application programming interfaces, and software development kits (Foster *et al.*, 2001). Computational Grids, with their attendant networks of people and instruments, are ideal for global-scale data mining and analysis (Craddock *et al.*, 2008).

Grid computing facilitates the use of workflows: analysis pipelines in which the output of one analysis feeds into the input of the next. Researchers have been carrying out this procedure manually since the emergence of the affordable computer, but current workflows can be completely automated. Automated workflows are built upon *Web services*.

Web services are formally defined interfaces that allow computational resources to be exposed in a standard, computationally comprehensible manner. Programs can be "exposed" as Web services by adding "wrapper" code, adhering to these standards, to the core program code. Web services may be hosted anywhere on the planet, and combined seamlessly into workflows—at least in theory. In practice, the use of workflows is fraught with practical difficulties relating to issues such as availability (the Web services upon which a workflow depends may go down without warning), reliability (the builder of a workflow cedes control of its components to their programmers, and the resulting code may or may not perform as intended) and documentation (many Web services perform excellently as designed, but many programmers are more focused upon the code than on documentation, making the Web service hard to use). Despite these drawbacks, well-designed workflows can perform tasks that would be prohibitive in terms of time and cost if carried out manually. Workflows also facilitate the automated re-analysis of data, as new datasets become available. Some applications, such as Microbase (Flanagan *et al.*, 2012), retain the results of previous analyses and process new data without the need to re-analyse the previously analysed data.

Several programs exist to facilitate the construction of fully automated workflows; examples are Taverna (Oinn *et al.*, 2004) and Microbase (Flanagan *et al.*, 2012). Workflows built using these tools can be stored and shared in repositories such as MyExperiment[16] (Goble *et al.*, 2010) (Figure 2.17).

Workflows have been applied to several large-scale problems, such as understanding the reaction of *E. coli* to oxygen (Maleki-Dizaji *et al.*, 2009); identification of microbial habitats (Kolluru *et al.*, 2011); analysis of structural differences in metabolic pathways (Arrigo *et al.*, 2007).

A relatively recent development, which makes it possible to perform unprecedentedly large amounts of computational analysis, is *Cloud* computing. Cloud computing includes "both the applications delivered as services over the Internet and the hardware and systems software in the data centres that provide those services"

---

[16]http://www.myexperiment.org/.

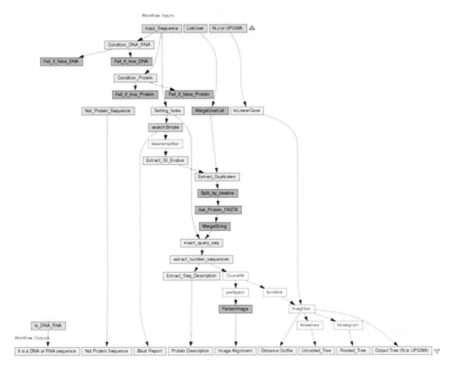

FIGURE 2.17

Taverna workflow for generic protein analysis. (For colour version of this figure, the reader is referred to the online version of this chapter.)

From the MyExperiment Web site (http://www.myexperiment.org/workflows/124/versions/1/previews/full) by M.B. Monteiro.

(Armbrust *et al.*, 2010). In practical terms, the maturing of Cloud computing, a long-held dream of the IT community, means that the analysis of large datasets is no longer constrained by the availability of compute resources at either the personal or institutional level. Several of the major online companies offer very affordable and reasonably readily learnt access to thousands of CPUs from any web-enabled desktop computer. Computational problems that would otherwise take thousands of hours to process using current computer technology can be farmed out to the Cloud, and completed in an hour or two using several thousand CPUs.

For example, the online seller Amazon offers access to Cloud facilities via the Amazon Elastic Compute Cloud (Amazon EC2).[17] A user can set up an account using a credit card, log in and access not only computers, but also a range of different software and hardware setups (known as *instances*). At the time of writing, a default on-demand instance was priced at $US0.115 per hour. Similarly, Google offers

---

[17]http://aws.amazon.com/ec2/.

Cloud facilities tailored to the users' needs.[18] Cloud computing means that data-intensive investigations such as genome-scale comparative and metagenomic analyses can be performed in a timely and cost-effective manner (Wilkening, 2009).

An innovative, recent approach to biomedical data mining, made possible by the Internet and the large amounts of Cloud storage space currently available, is crowdsourcing. Crowdsourcing is "the practice of obtaining needed services, ideas or content by soliciting contributions from a large group of people and especially from the online community rather than from traditional employees or suppliers."[19] This type of approach has been applied to scientific problems ever since the Internet became ubiquitous; perhaps the most well-known examples are the Search for Extraterrestrial Intelligence (SETI)[20] and Folding@Home.[21]

The SETI Institute was founded in 1984, and uses crowdsourcing to examine radio frequency signals from the SETI Institute's Allen Telescope Array for indications of possible alien civilizations. Users donate spare domestic CPU cycles to run the analysis software. Folding@home is more biologically oriented, using volunteers' computers to run protein-folding simulations. Both of these projects rely more upon users providing computational power than intellectual power, but as computation becomes increasingly cost-effective, the focus has turned to true crowdsourcing. In the field of astronomy, Galaxy Zoo[22] uses volunteers to classify galaxies in images from a number of telescopes, including the Hubble Space Telescope, in order to investigate how galaxies form and evolve. Over 60 million classifications have been made to date, and overall the classifications are as good as those made by professional astronomers (Lintott et al., 2008).

Crowdsourcing does not appear to have been widely used in microbiology until recently. In 2011 Germany experienced an outbreak of haemolytic uraemic syndrome with bloody diarrhoea. It was caused by the virulent *E. coli* strain O104:H4. By the time the organism was identified and sequenced 845 cases and 54 deaths had occurred (Bielaszewska et al., 2011). *Nature Biotechnology* described it as "the most deadly *E. coli* outbreak on record" (Outbreak Genomics [Editorial], 2011). Researchers from the Beijing Genomics Institute sequenced the organism on an Ion Torrent Personal Genome Machine,[23] and made the data freely available on the github Web site.[24] Bioinformaticists all around the world tackled the data; a *de novo* assembly was completed the following day, and the full annotation of the draft genome was completed within a week (Rohde et al., 2011).

During this time additional data became available as a number of different centres re-sequenced using a range of "Next Generation Sequencing" platforms. Although

---

[18]http://cloud.google.com/.
[19]http://www.merriam-webster.com/dictionary/crowdsourcing.
[20]www.seti.org.
[21]http://folding.stanford.edu/English/HomePage.
[22]http://www.galaxyzoo.org/.
[23]Life Technologies www.lifetechnologies.com.
[24]https://github.com.

the rapid annotation of this genome did not affect the clinical treatment of the outbreak, it was a convincing demonstration of the potential application of microbial data mining via the wisdom of crowds.

> **Software Availability**
>
> Taverna: http://www.taverna.org.uk/.
>
> Microbase: http://www.microbasecloud.com/.

7 CASE STUDY: DATA MINING FOR PROTEIN FUNCTION PREDICTION

One of the major issues challenging bioinformatics in general, and microbial genetics in particular, is the prediction of protein function from DNA sequence. The number of fully sequenced microbial genomes is growing exponentially, and the advent of Next Generation Sequencing technologies means that the rate at which new genomes are acquired will increase inexorably. However, the percentage of reliably annotated proteins drops proportionately to the number of genomes sequenced, since the process of experimentally producing such annotations takes far more time than does the generation of sequence.

The workhorse of protein functional prediction is still BLAST. New sequences are compared with sequences already on record, and if two sequences are similar enough, annotations may be transferred from one to another. Although the basic assumption—that similar sequences are likely to produce proteins with similar function—is broadly supportable, this assumption does not always hold. Further, over time, this practice leads to a phenomenon colloquially known as *database rot*. Sequence A might be very similar to sequence B, which was annotated on the basis of its similarity to sequence C, which was annotated... and so on. Although sequence A might be very similar to sequence B, it may be quite different from sequence Z, the originally, experimentally verified protein. The annotation of sequence A may therefore be far from accurate.

A large number of approaches have been taken to protein functional assignment; a good review is provided by (Sleator, 2012). In this review we are concerned only with those methods that can broadly be described as data mining (as opposed to, for example, deductions from measures of evolutionary relatedness). As with many bioinformatics tasks, multiple algorithms are frequently combined in order to address this problem.

Several research groups have adopted the approach of calculating distances between proteins, which are then used to build classification or decision trees, which are then turned into classification rules.

An early example of this approach was learning of classification rules to infer protein function in *Mycobacterium tuberculosis* and *E. coli* (King et al., 2000). The process started with building a decision tree, using C4.5 and C5.0 (see above). Inductive Logic Programming (ILP) was then used to derive rules from the decision trees, which were then pruned to avoid overfitting. ILP uses the language of logic programs to describe examples and theories, and is as powerful and flexible as general-purpose programming languages such as Java. The ILP approach used was aimed at identifying frequent patterns in the data and using these patterns to identify clusters. The clusters were then converted into rules of the form:

IF A THEN B

The functions of 65% of the ORFs in *M. tuberculosis* and 24% of those in *E. coli* were predicted with 60–80% accuracy using this approach. The same group used mutant phenotype growth data to predict the functional class of ORFs in *S. cerevisiae*, again using a modified version of C4.5 to produce classification rules (Clare and King, 2002).

A similar approach used two different decision tree algorithms to learn rules for the annotation of proteins from *Lactobacillus sakei* with terms from a controlled vocabulary developed originally for *B. subtilis* (Moszer *et al.*, 2002). The rules were learned from a training set of data for *L. bulgaricus* (Azé *et al.*, 2007). It achieved a precision of 80.5%, with a recall of 52.7%.

Interactomes have been widely used for the prediction of protein function. The approaches taken usually involve constructing a network, clustering it, and then using cluster membership to predict protein function. There are many algorithms, ranging from simply taking the most-frequent annotation amongst the neighbours of a protein (Schwikowski *et al.*, 2000), to statistically based probabilistic methods (Letovsky, 2003; Joshi *et al.*, 2005; Kao and Huang, 2010), and graph-theoretic methods (Nabieva *et al.*, 2005). Machine learning techniques applied to interactomes include the calculation of Bayesian likelihood scores using homology information from other genomes (Date and Stoeckert, 2006), Markovian random field theory (Letovsky, 2003; Deng *et al.*, 2004a,b) and Bayesian approaches (Jansen *et al.*, 2003; Troyanskaya *et al.*, 2003; Nariai *et al.*, 2007)

One of the most effective approaches, however, turns out to be one of the simplest: propagation of functional labels to an un-annotated protein via the neighbour with the highest weight, either level 1 neighbours, level 2 neighbours, or both (Chua *et al.*, 2006, 2007)

An approach that combines interactome analysis with clustering, classification tree construction and rule inference, was described by Brun and colleagues (Brun *et al.*, 2004). These authors achieved an accuracy of between 58% and 64%, depending upon the subset of data considered.

Data mining has also been applied successfully to the prediction of protein function in metagenomic datasets. Such datasets are challenging because they generally come from a wide variety of organisms with different characteristics, such as GC content and codon usage bias. The majority of the organisms identified in most metagenomic studies are almost completely un-annotated, since more than 99% of prokaryotes in the environment cannot be cultured in the laboratory (Schloss and Handelsman, 2005). BLAST searches alone can therefore only provide limited information about protein function.

Several groups have applied clustering to metagenomics data. When metagenomic data are clustered, proteins can be grouped together either on the basis of their domains (Corpet *et al.*, 1998; Bateman *et al.*, 2004) or their full sequences (Haft *et al.*, 2003; Yooseph *et al.*, 2007). Filtering can then be applied to the clusters to eliminate high-distance links within clusters and detect spurious ORFs (Yooseph *et al.*, 2008). Protein function prediction is then achieved using a guilt-by-association approach.

8 DATA MINING WITH MICROBIAL DATA: PRACTICAL ISSUES

There is an enormous amount of microbiological data already in existence, with more constantly generated by ever-improving and expanding technologies. Data mining clearly has the potential to identify significant trends and patterns in this data. However, biological data in general, and microbiological data in particular, pose particular problems for data miners.

8.1 Noise

Biological data is inherently noisy. Biological noise arises from a wide range of sources, not all of which are understood. Biological noise may be *intrinsic*, due to stochasticity in processes such as transcription and translation, or *extrinsic*, arising from fluctuations in the environment (Elowitz, 2002; Swain *et al.*, 2002).

Intrinsic noise can arise in gene regulation, for example, when the chance of a TF binding depends upon the number of TF molecules in the cell, meaning transcription tends to occur in bursts, rather than as a consistent, predictable process (McAdams and Arkin, 1997). More than 80% of the genes in the *E. coli* chromosome express fewer than a hundred copies each of their protein products per cell (Guptasarma, 1995). The rates of transcription, translation, modification or degradation of RNAs and proteins vary for different gene products (Newman *et al.*, 2006). The number of mRNA molecules in a cell is thus variable, even under the same environmental and genetic conditions; and of course, genetically identical cells are exposed to subtly different extrinsic factors—micro-fluctuations in temperature, pH, nutrient availability and crowding, even under apparently identical experimental conditions.

Implicit in the analysis of most microarray data is the general assumption that mRNA levels are directly correlated with protein levels; however, this is not always the case. These factors, and many others, lead to variability in the numbers of the specific biomolecules that are physically present in cells. In addition to biological sources, measurement processes introduce variability into data. No laboratory equipment, including the human eye and brain, performs with 100% accuracy.

Microbes deal with noise either by overcoming its effects (You *et al.*, 2004; Austin *et al.*, 2006), or by incorporating it into day-to-day life (Ross *et al.*, 1994; Maheshri and O'Shea, 2007). Consequently, when a data miner considers a biological dataset, it is never clear to what extent the noise is important. However, many algorithms are affected by the presence of noise in a dataset, and these effects must be borne in mind during the analysis.

8.2 Overfitting

One of the issues that often arises is overfitting. Overfitting occurs when an algorithm that is trained using labelled data learns the characteristics of the training data too well, to the point where its performance upon previously unseen data deteriorates.

Many CI-oriented data mining algorithms are prone to overfitting. All datasets contain both signals—the patterns in the data that are important—and noise—errors due to random chance or variation in equipment performance. The aim of a data mining algorithm is to learn the characteristics of the signal. However, many algorithms will learn the noise as well, particularly if the training dataset is small.

Noise, due to its random nature, will be unique to a particular training set, and overfitting thus leads to a decrease in the performance of the algorithm on unseen data (Figure 2.18). Noise can be reduced, for example, by using multiple forms of measurement, but can never be completely eliminated.

To avoid overfitting it is important to make the most effective use of the available data. If enough data are available, the ideal situation is to have three completely separate datasets: training, validation and test. The training dataset is used, on its own, to train the algorithm. The performance of the trained algorithm is then assessed using the previously unseen validation dataset. If the performance of the algorithm is not adequate, changes can be made to the training set, and the algorithm re-trained and re-validated. By using the validation set in this way, however, it essentially becomes

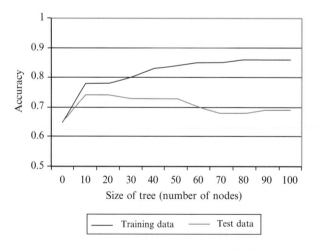

FIGURE 2.18

Overfitting leads to decreased performance of an algorithm on unseen data. In this case performance is recorded for a set of decision trees with varying numbers of nodes. (For colour version of this figure, the reader is referred to the online version of this chapter.)

part of the training process. Therefore, the performance of the algorithm should be reported as that achieved on the completely unseen test set.

If there are insufficient data to establish three separate datasets, the next-most-parsimonious approach is the use of cross-validation. With cross-validation the data is divided into a number of datasets, often either three or ten, depending upon the amount of data available. One dataset is held out, and the algorithm trained on the rest. The held-out data is then classified using the trained classifier. This process is repeated with each held-out dataset in turn. The end result of this approach is that, in every case, the data is classified by a classifier on which it was not trained. The trade-off is that, since the classifiers are trained upon smaller subsets of data, they are likely to perform less well than a classifier trained upon the entire dataset. The ultimate form of cross-validation is the "leave-one-out cross-validation" approach, in which each case is individually held-out in turn (Witten et al., 2011).

8.3 The peaking phenomenon

It appears to be intuitively obvious that providing more descriptive variables to a data mining algorithm will improve its performance. To some extent this assumption is valid. However, all data contain noise as well as the specific signal, particularly that generated by high-throughput approaches. Eventually, the addition of new variables will actually degrade rather than enhance an algorithm's performance, a scenario known as the *peaking phenomenon* (Figure 2.19) (Sima and Dougherty, 2008). The peaking phenomenon does not always occur, but should be tested for by running the algorithms on different-sized subsets of the available data to check the effect, on the accuracy of the output, of the number of variables included in the analysis.

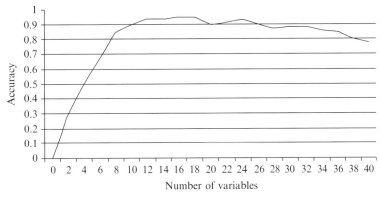

FIGURE 2.19

The peaking phenomenon. As variables are added to an analysis, the accuracy of the classification initially rises. Eventually, the addition of more variables introduces more noise than signal, and the performance of the algorithm deteriorates.

9 CONCLUSIONS

The field of data mining encompasses a huge range of techniques, addressing a very wide range of problems. Standard statistical approaches have much to offer, if the data are appropriate and the algorithms computationally feasible. With molecular microbiological data, however, this is not always the case. Many high-throughput datasets are large, complex and noisy, with unknown degrees of error. Despite these drawbacks, there are many CI-based algorithms that can be used to deduce new information from these datasets. Data mining algorithms must, however, always be used with caution, and their results scrutinised carefully by knowledgeable experts. In many cases, the most appropriate use of data mining is to inspire new, testable hypotheses based upon previously unseen patterns in datasets.

The rise in the size of microbiological datasets generated by new, high-throughput technologies is excitingly mirrored by the advent of new computational and social paradigms for the analysis of large datasets. Classic algorithms for clustering, classification and comparison can be applied to new datasets of unprecedented size using Grid and Cloud technologies, and the interpretation of results can, at the discretion of the researcher, be made available to millions of crowd-sourced minds. Data mining first arose decades ago as a means by which marketers could make the most of trends in consumer behaviour, but it currently offers the promise of guiding microbiologists towards a deeper understanding of microbial behaviour.

References

Abdi, H. and Williams, L. J. (2010). Principal component analysis. *WIRES Comput. Stat.* **2**, 433–459.

Altschul, S. F., Gish, W., Miller, W., Myers, E. W., and Lipman, D. J. (1990). Basic local alignment search tool. *J. Mol. Biol.* **215**, 403–410.

Altschul, S. F., Madden, T. L., Schäffer, A. A., Zhang, J., Zhang, Z., Miller, W., and Lipman, D. J. (1997). Gapped BLAST and PSI-BLAST: a new generation of protein database search programs. *Nucleic Acids Res.* **25**, 3389–3402.

Altschuler, D., Daly, M., and Kruglyak, L. (2000). Guilt by association. *Nat. Genet.* **26**, 135–137.

Armbrust, M., Fox, A., Griffith, R., Joseph, A. D., Katz, R., Konwinski, A., Lee, G., Patterson, D., Rabkin, A., Stoica, I., and Zaharia, M. (2010). A view of Cloud computing. *Commun. ACM* **53**, 50–58.

Arrigo, P., Cardo, P. P., and Ruggiero, C. (2007). Integrated bioinformatics analysis of structural differences in metabolic pathways. An application to *Mycobacterium leprae*. In: *2nd IEEE International Conference on Nano/Micro Engineered and Molecular Systems, 2007 Bangkok, Thailand*.

Ashburner, M., Ball, C. A., Blake, J. A., Botstein, D., Butler, H., Cherry, J. M., Davis, A. P., Dolinski, K., Dwight, S. S., Eppig, J. T., Harris, M. A., Hill, D. P., Issel-Tarver, L., Kasarskis, A., Lewis, S., Matese, J. C., Richardson, J. E., Ringwald, M., Rubin, G. M., and Sherlock, G. (2000). Gene Ontology: tool for the unification of biology. *Nat. Genet.* **25**, 25–29.

Austin, D. W., Allen, M. S., McCollum, J. M., Dar, R. D., Wilgus, J. R., Sayler, G. S., Samatova, N. F., Cox, C. D., and Simpson, M. L. (2006). Gene network shaping of inherent noise spectra. *Nature* **439**, 608–611.

Azad, R. K. and Borodovsky, M. (2004). Probabilistic methods of identifying genes in prokaryotic genomes: connections to the HMM theory. *Brief. Bioinform.* **5**, 118–130.

Azé, J., Gentils, C., Toffano-Nioche, C., Loux, V., Gibrat, J.-F., Bessières, P., Rouveirol, C., Poupon, A., and Froidevaux, C. (2007). Towards a semi-automatic functional annotation tool based on decision-tree techniques. *BMC Proc.* **2**, S3.

Bachmann, H., Starrenburg, M. J. C., Dijkstra, A., Molenaar, D., Kleerebezem, M., Rademaker, J. L. W., and Van Hylckama Vlieg, J. E. T. (2009). Regulatory phenotyping reveals important diversity within the species Lactococcus lactis. *Appl. Environ. Microbiol.* **75**, 5687–5694.

Ballesté, E., Bonjoch, X., Belanche, L. A., and Blanch, A. R. (2010). Molecular indicators used in the development of predictive models for microbial source tracking. *Appl. Environ. Microbiol.* **76**, 1789–1795.

Bateman, A., Coin, L., Durbin, R., Finn, R. D., Hollich, V., Griffiths-Jones, S., Khanna, A., Marshall, M., Moxon, S., Sonnhammer, E. L., Studholme, D. J., Yeats, C., and Eddy, S. R. (2004). The Pfam protein families database. *Nucleic Acids Res.* **32**, D138–D141.

Baum, L. E. and Petrie, T. (1966). Statistical inference for probabilistic functions of finite state Markov chains. *Annals Math. Stat.* **37**, 1554–1563.

Belanche-Muñoz, L. and Blanch, A. R. (2008). Machine learning methods for microbial source tracking. *Environ. Model. Software* **23**, 741–750.

Besemer, J. and Borodovsky, M. (2005). GeneMark: web software for gene finding in prokaryotes, eukaryotes and viruses. *Nucleic Acids Res.* **33**, W451–W454.

Besemer, J., Lomsadze, A., and Borodovsky, M. (2001). GeneMarkS: a self-training method for prediction of gene starts in microbial genomes. Implications for finding sequence motifs in regulatory regions. *Nucleic Acids Res.* **29**, 2607–2618.

Bhasin, M., Garg, A., and Raghava, G. P. S. (2005). PSLpred: prediction of subcellular localization of bacterial proteins. *Bioinformatics* **21**, 2522–2524.

Bi, D., Xu, Z., Harrison, E. M., Tai, C., Wei, Y., He, X., Jia, S., Deng, Z., Rajkumar, K., and Ou, H.-K. (2012). ICEberg: a web-based resource for integrative and conjugative elements found in Bacteria. *Nucleic Acids Res.* **40**, D621–D626.

Bielaszewska, M., Mellmann, A., Zhang, W., Köck, R., Ruth, A., Bauwens, A., Peters, G., and Karch, H. (2011). Characterisation of the *Escherichia coli* strain associated with an outbreak of haemolytic uraemic syndrome in Germany, 2011: a microbiological study. *Lancet Infect. Dis.* **11**, 671–676.

Bishop, C. M. (2007). *Pattern Recognition and Machine Learning.* Springer.

Blackwood, C. B., Marsh, T., Kim, S.-H., and Paul, E. A. (2003). Terminal restriction fragment length polymorphism data analysis for the quantitative comparison of microbial communities. *Appl. Environ. Microbiol.* **69**, 926.

Boser, B. E., Guyon, I. M., and Vapnik, V. N. (1992). A training algorithm for optimal margin classifiers. In: D. Haussler (Ed.), *5th Annual ACM Workshop on COLT*, ACM Press.

Boutros, P. C. and Okey, A. B. (2005). Unsupervised pattern recognition: An introduction to the whys and wherefores of clustering microarray data. *Brief. Bioinform.* **6**(4), 331–343.

Brun, C., Chevenet, F., Martin, D., Wojcik, J., Guénoche, A., and Jacq, B. (2004). Functional classification of proteins for the prediction of cellular function from a protein-protein interaction network. *Genome Biol.* **5**, R6.

Byrne, D. and Uprichard, E. (2012). *Cluster Analysis.* UK: SAGE Publications Ltd.

Cai, C. Z., Han, L. Y., Ji, Z. L., Chen, X., and Chen, Y. Z. (2003). SVM-Prot: web-based support vector machine software for functional classification of a protein from its primary sequence. *Nucleic Acids Res.* **31**, 3692–3697.

Ceci, F., Pietrobon, R., and Gonçalves, A. L. (2012). Turning text into research networks: information retrieval and computational ontologies in the creation of scientific databases. *PLoS One* **7**, e27499.

Chua, H. N., Sung, W.-K., and Wong, L. (2006). Exploiting indirect neighbours and topological weight to predict protein function from protein–protein interactions. *Bioinformatics* **22**, 1623–1630.

Chua, H. N., Sung, W.-K., and Wong, L. (2007). Using indirect protein interactions for the prediction of Gene Ontology functions. *BMC Bioinformatics* **8**, S8.

Clare, A. and King, R. D. (2002). Machine learning of functional class from phenotype data. *Bioinformatics* **18**, 160–166.

Clermont, O., Bonacorsi, S., and Bingen, E. (2000). Rapid and simple determination of the *Escherichia coli* phylogenetic group. *Appl. Environ. Microbiol.* **66**, 4555–4558.

Coenen, F. (2011). Data mining: past, present and future. *Knowl. Eng. Rev.* **26**, 25–29.

Corpet, F., Gouzy, J., and Kahn, D. (1998). The ProDom database of protein domain families. *Nucleic Acids Res.* **26**, 323–326.

Craddock, T., Harwood, C. R., Hallinan, J., and Wipat, A. (2008). e-Science: relieving bottlenecks in large-scale genomic analyses. *Nat. Rev. Microbiol.* **6**, 948–954.

Cummings, C. A. and Relman, D. A. (2000). Using DNA microarrays to study host-microbe interactions. *Emerg. Infect. Dis.* **6**, 513–525.

Cusick, M. E., Yu, H., Smolyar, A., Venkatesan, K., Carvunis, A.-R., Simonis, N., Rual, J.-F., Borick, H., Braun, P., Dreze, M., Vandenhaute, J., Galli, M., Yazaki, J., Hill, D. E., Ecker, J. R., Roth, F. P., and Vidal, M. (2008). Literature-curated protein interaction datasets. *Nat. Methods* **6**, 39–46.

Date, S. V. and Stoeckert, C. J. (2006). Computational modeling of the *Plasmodium falciparum* interactome reveals protein function on a genome-wide scale. *Genome Res.* **16**, 542–549.

Delmont, T. O., Robe, P., Cecillon, S., Clark, I. M., Constancias, F., Simonet, P., Hirsch, P. R., and Vogel, T. M. (2011). Accessing the soil metagenome for studies of microbial diversity. *Appl. Environ. Microbiol.* **77**, 1315–1324.

Deng, M., Chen, T., and Sun, F. (2004a). An integrated probabilistic model for functional prediction of proteins. *J. Comput. Biol.* **11**, 463–475.

Deng, M., Tu, Z., Sun, F., and Chen, T. (2004b). Mapping gene ontology to proteins based on protein–protein interaction data. *Bioinformatics* **20**, 895–902.

Dieckmann, R. and Malorny, B. (2011). Rapid screening of epidemiologically important *Salmonella enterica* subsp. *enterica* serovars by whole-cell Matrix-Assisted Laser Desorption Ionization–Time of Flight mass spectrometry. *Appl. Environ. Microbiol.* **77**, 4136–4146.

Dobrin, R., Beq, Q. C., Barabasi, A. L., and Oltvai, S. N. (2004). Aggregation of topological motifs in the *Eschericia coli* transcriptional regulatory network. *BMC Bioinformatics* **5**, 10.

Elowitz, M. B. (2002). Stochastic gene expression in a single cell. *Science* **297**, 1183–1186.

Evans, J. A. and Rzhetsky, A. (2011). Advancing science through mining libraries, ontologies, and communities. *J. Biol. Chem.* **286**, 23659–23666.

Faris, J., Kolker, E., Szalay, A., Bradlow, L., Deelman, E., Feng, W., Qiu, J., Russell, D., Stewart, E., and Kolker, E. (2011). Communication and data-intensive science in the beginning of the 21st century. *OMICS J. Integr. Biol.* **15**, 213–215.

Fentress, G. L. (1914). Tendencies in modern educational development; Annual address delivered before the alumni association. *Bulletin* **7**, 29.

Ferdinand, S., Valétudie, G., Sola, C., and Rastogi, N. (2004). Data mining of *Mycobacterium tuberculosis* complex genotyping results using mycobacterial interspersed repetitive units validates the clonal structure of spoligotyping-defined families. *Res. Microbiol.* **155**, 647–654.

Fernández-Navarro, F., Valero, A., Hervás-Martínez, C., Gutiérrez, P. A., García-Gimeno, R. M., and Zurera-Cosano, G. (2010). Development of a multi-classification neural network model to determine the microbial growth/no growth interface. *Int. J. Food Microbiol.* **141**, 203–212.

Fields, S. and Song, O.-K. (1989). A novel genetic system to detect protein–protein interactions. *Nature* **340**, 245–246.

Findlen, P. (2004). *Athanasius Kircher: The Last Man Who Knew Everything*. Routledge.

Flanagan, K., Nakjang, S., Hallinan, J., Harwood, C., Hirt, R. P., Pocock, M. R., and Wipat, A. (2012). Microbase2.0: a generic framework for computationally intensive bioinformatics workflows in the cloud. In: *2012 International Symposium on Integrative Bioinformatics (IB2012). Hangzhou, China*.

Forney, G.Jr., (1973). The Viterbi algorithm. *Proc. IEEE* **61**, 268–278.

Foster, I., Kesselman, C., and Tuecke, S. (2001). The anatomy of the Grid: enabling scalable virtual organizations. *Int. J. High Perform. Comput. Appl.* **15**, 200–222.

Foster, J. A., Bunge, J., Gilbert, J. A., and Moore, J. H. (2012). Measuring the microbiome: perspectives on advances in DNA-based techniques for exploring microbial life. *Brief. Bioinform.* **13**, 420–429.

Frank, D. N., St. Amand, A. L., Feldman, R. A., Boedeker, E. C., Harpaz, N., and Pace, N. (2007). Molecular-phylogenetic characterization of microbial community imbalances in human inflammatory bowel diseases. *Proc. Natl. Acad. Sci.* **104**, 13780–13785.

Friedman, J. H. (1991). Multivariate adaptive regressive splines. *Annals Stat.* **19**, 1–67.

Fukunaga, K. and Narendra, P. M. (1975). A branch and bound algorithm for computing k-nearest neighbors. *IEEE Trans. Comput.* **1975**, 750.

Galperin, M. Y. and Fernández-Suárez, X. M. (2012). The 2012 nucleic acids research database issue and the online molecular biology database collection. *Nucleic Acids Res.* **40**, D1–D8.

Gardy, J. L. and Brinkman, F. S. L. (2006). Methods for predicting bacterial protein subcellular localization. *Nat. Rev. Microbiol.* **4**, 741–751.

Gardy, J. L., Laird, M. R., Chen, F., Rey, S., Walsh, C. J., Ester, M., and Brinkman, F. S. L. (2005). PSORTb v.2.0: expanded prediction of bacterial protein subcellular localization and insights gained from comparative proteome analysis. *Bioinformatics* **21**, 617–623.

Gelfand, M. (2006). Evolution of transcriptional regulatory networks in microbial genomes. *Curr. Opin. Struct. Biol.* **16**, 420–429.

Goble, C. A., Bhagat, J., Aleksejevs, S., Cruickshank, D., Michaelides, D., Newman, D., Borkum, M., Bechhofer, S., Roos, M., Li, P., and De Roure, D. (2010). myExperiment: a repository and social network for the sharing of bioinformatics workflows. *Nucleic Acids Res.* **38**, W677–W682.

Goffeau, A., Barrell, B. G., Bussey, H., Davis, R. W., Dujon, B., Feldmann, H., Galibert, F., Hoheisel, J. D., Jacq, C., Johnston, M., Louis, E. J., Mewes, H. W., Murakami, Y., Philippsen, P., Tettelin, H., and Oliver, S. G. (1996). Life with 6000 Genes. *Science* **274**, 546–567.

Goll, J., Rajagopala, S. V., Shiau, S. C., Wu, H., Lamb, B. T., and Uetz, P. (2008). MPIDB: the microbial protein interaction database. *Bioinformatics* **24**, 1743–1744.

Gruber, T. R. (1993). A translation approach to portable ontology specifications. *Knowl. Acqu.* **5**, 199–220.

Guptasarma, P. (1995). Does replication-induced transcription regulate synthesis of the myriad low copy number proteins of *Escherichia coli*? *Bioessays* **17**, 987–997.

Gutenkunst, R. N., Waterfall, J. J., Casey, F. P., Brown, K. S., Myers, C. R., and Sethna, J. P. (2007). Universally sloppy parameter sensitivities in systems biology models. *PLoS Comput. Biol.* **3**, e189.

Haft, D. H., Selengut, J. D., and White, O. (2003). The TIGRFAMs database of protein families. *Nucleic Acids Res.* **31**, 371–373.

Hallinan, J. and Wipat, A. (2006). Clustering and crosstalk in a yeast functional interaction network. In: *2006 IEEE Symposium on Computational Intelligence in Bioinformatics and Computational Biology (CIBCB 2006)*, IEEE Press.

Hallinan, J., Pocock, M., Addinall, S. G., Lydall, D., and Wipat, A. (2009). Clustering incorporating shortest paths identifies relevant modules in functional interaction networks. In: *2009 IEEE Symposium on Computational Intelligence in Bioinformatics and Computational Biology (CIBCB09)*.

Hallinan, J., James, K., and Wipat, A. (2011). Network approaches to the functional analysis of microbial proteins. *Adv. Microb. Physiol.* **59**, 101–123.

Hamming, R. W. (1950). Error detecting and error correcting codes. *Bell System Technical Journal* **29**, 147–160.

Hartigan, J. A. and Wong, M. A. (1979). Algorithm AS 136: a k-means clustering algorithm. *J. R. Stat. Soc. Ser. C (Applied Statistics)* **28**, 100–108.

Haykin, S. O. (2008). *Neural Networks and Learning Machines: (3rd edition) A Comprehensive Foundation*. Pearson: New York.

Hecht-Nielsen, R. (1989). Theory of the backpropagation neural network. In: *International Joint Conference on Neural Networks (IJCNN)*. San Diego, CA.

Herrgård, M. (2004). Reconstruction of microbial transcriptional regulatory networks. *Curr. Opin. Biotechnol.* **15**, 70–77.

Hu, X., Zhang, X., Wu, D., Zhou, X., and Rumm, P. (2008). Text mining the biomedical literature for identification of potential virus/bacterium as bio-terrorism weapons. *Terrorism Inf.* **18**, 385–406 Integrated Series in Information Systems.

Hunter, L. and Cohen, K. B. (2006). Biomedical language processing: what's beyond PubMed? *Mol. Cell* **21**, 589–594.

Huse, S. M., Welch, D. M., Morrison, H. G., and Sogin, M. L. (2010). Ironing out the wrinkles in the rare biosphere through improved OTU clustering. *Environ. Microbiol.* **12**, 1889–1898.

Illian, J. B., Prosser, J. I., Baker, K. L., and Rangel-Castro, J. I. (2009). Functional principal component data analysis: a new method for analysing microbial community fingerprints. *J. Microbiol. Methods* **79**, 89–95.

James, K., Wipat, A., and Hallinan, J. (2009). Integration of full-coverage probabilistic functional networks with relevance to specific biological processes. In: *Data Integration in the Life Sciences (DILS2009)*, Springer **5647**, 31–46.

Janga, S. C. and Collado-Vides, J. (2007). Structure and evolution of gene regulatory networks in microbial genomes. *Res. Microbiol.* **158**, 787–794.

Jansen, R., Yu, H., Greenbaum, D., Kluger, Y., Krogan, N. J., Chung, S., Emili, A., Snyder, M., Greenblatt, J. F., and Gerstein, M. (2003). A Bayesian networks approach for predicting protein-protein interactions from genomic data. *Science* **302**, 449–453.

Jensen, L. J. and Bork, P. (2010). Ontologies in quantitative biology: a basis for comparison, integration, and discovery. *PLoS Biol.* **8**, e1000374.

Joshi, T., Chen, Y., Becker, J. M., Alexandrov, N., and Xu, D. (2005). Genome-scale gene function prediction using multiple sources of high-throughput data in yeast *Saccharomyces cerevisiae*. *OMICS J. Int. Biol.* **8**, 322–333.

Juck, D., Charles, T., Whyte, L. G., and Greer, C. W. (2000). Polyphasic microbial community analysis of petroleum hydrocarbon-contaminated soils from two northern Canadian communities. *FEMS Microbiol. Ecol.* **33**, 241–249.

Kao, K.-C. and Huang, J.-Y. (2010). Accurate and fast computational method for identifying protein function using protein-protein interaction data. *Mol. Biosyst.* **6**, 830–839.

Kass, G. V. (1980). An exploratory technique for exploring large quantities of data. *Appl. Stat.* **29**, 119–127.

Keim, P., Price, L. B., Klevytska, A. M., Smith, K. L., Schupp, J. M., Okinaka, R., Jackson, P. J., and Hugh-Jones, M. E. (2000). Multiple-locus variable-number tandem repeat analysis reveals genetic relationships within *Bacillus anthracis*. *J. Bacteriol.* **182**, 2928–2936.

Kennedy, R. L. (1997). *Solving Data Mining Problems Through Pattern Recognition*. Prentice Hall PTR.

Kesselman, C. and Foster, I. (1998). *The Grid: Blueprint for a New Computing Infrastructure*. Morgan Kaufmann Publishers.

King, R. D., Karwath, A., Clare, A., and Dehaspe, L. (2000). Accurate prediction of protein functional class from sequence in the *Mycobacterium tuberculosis* and *Escherichia coli* genomes using data mining. *Yeast* **17**, 283–293.

Kolluru, B., Nakjang, S., Hirt, R. P., Wipat, A., and Ananiadou, S. (2011). Automatic extraction of microorganisms and their habitats from free text using text mining workflows. *J. Integr. Bioinform.* **8**, 184–194.

Krause, L., Mchardy, A. C., Nattkemper, T. W., Pühler, A., Stoye, J., and Meyer, F. (2007). GISMO—gene identification using a support vector machine for ORF classification. *Nucleic Acids Res.* **35**, 540–549.

Krogan, N. J., Cagney, G., Yu, H., Zhong, G., Guo, X., Ignatchenko, A., Li, J., Pu, S., Datta, N., Tikuisis, A. P., Punna, T., Peregrín-Alvarez, J. M., Shales, M., Zhang, X., Davey, M., Robinson, M. D., Paccanaro, A., Bray, J. E., Sheung, A., Beattie, B., Richards, D. P., Canadien, V., Lalev, A., Mena, F., Wong, P., Starostine, A., Canete, M. M., Vlasblom, J., Wu, S., Orsi, C., Collins, S. R., Chandran, S., Haw, R., Rilstone, J. J., Gandi, K., Thompson, N. J., Musso, G., St. Onge, P., Ghanny, S., Lam, M. H. Y., Butland, G., Altaf-Ul, A. M., Kanaya, S., Shilatifard, A., O'Shea, E., Weissman, J. S., Ingles, C. J., Hughes, T. R., Parkinson, J., Gerstein, M., Wodak, S. J., Emili, A., and Greenblatt, J. F. (2006). Global landscape of protein complexes in the yeast *Saccharomyces cerevisiae*. *Nature* **440**, 637–643.

Krogh, A. (1998). Chapter 4 an introduction to hidden Markov models for biological sequences. In S. L. Salzberg, D. B. Searls & S. Kasif (Eds.), *New Comprehensive Biochemistry*. Elsevier.

Kumar, M., Verma, R., and Raghava, G. P. S. (2006). Prediction of mitochondrial proteins using Support Vector Machine and Hidden Markov Model. *J. Biol. Chem.* **281**, 5357–5363.

Langille, M. G. I., Hsiao, W. W. L., and Brinkman, F. S. L. (2010). Detecting genomic islands using bioinformatics approaches. *Nat. Rev. Microbiol.* **8**, 373–382.

Larsen, P. E., Field, D., and Gilbert, J. A. (2012). Predicting bacterial community assemblages using an artificial neural network approach. *Nat. Methods* **9**, 621–625.

Lee, I., Date, S., Adai, A., and Marcotte, E. (2004a). A probabilistic functional network of yeast genes. *Science* **306**, 1555–1558.

Lee, Y., Lin, Y., and Wahba, G. (2004b). Multicategory support vector machines. *J. Am. Stat. Assoc.* **99**, 67–81.

Lemos, L. N., Fulthorpe, R. R., Triplett, E. W., and Roesch, L. F. W. (2011). Rethinking microbial diversity analysis in the high throughput sequencing era. *J. Microbiol. Methods* **86**, 42–51.

Letovsky, S. (2003). Predicting protein function from protein/protein interaction data: a probabilistic approach. *Bioinformatics* **19**, 197i–204i.

Leyfer, D. (2005). Genome-wide decoding of hierarchical modular structure of transcriptional regulation by cis-element and expression clustering. *Bioinformatics* **21**, ii197–ii203.

Lintott, C. J., Schawinski, K., Slosar, A., Land, K., Bamford, S., Thomas, D., Raddick, M. J., Nichol, R. C., Szalay, A., Andreescu, D., Murray, P., and Van Den Berg, J. (2008). Galaxy Zoo: morphologies derived from visual inspection of galaxies from the Sloan Digital Sky Survey. *MNRAS* **389**, 1179–1189.

Lu, Z. (2011). PubMed and beyond: a survey of web tools for searching biomedical literature. *J. Biol. Databases Curat.* **2011**, baq036.

Luciano, J. S. and Stevens, R. D. (2007). e-Science and biological pathway semantics. *BMC Bioinformatics* **8**, S3.

Lukashin, A. V. and Borodovsky, M. (1998). GeneMark.hmm: new solutions for gene finding. *Nucleic Acids Res.* **26**, 1107–1115.

Lyautey, E., Lapen, D. R., Wilkes, G., McCleary, K., Pagotto, F., Tyler, K., Hartmann, A., Piveteau, P., Rieu, A., Robertson, W. J., Medeiros, D. T., Edge, T. A., Gannon, V., and Topp, E. (2007). Distribution and characteristics of *Listeria monocytogenes* isolates from surface waters of the South Nation River watershed, Ontario, Canada. *Appl. Environ. Microbiol.* **73**, 5401–5410.

Lyautey, E., Lu, Z., Lapen, D. R., Wilkes, G., Scott, A., Berkers, T., Edge, T. A., and Topp, E. (2010). Distribution and diversity of *Escherichia coli* populations in the South Nation River drainage basin, Eastern Ontario, Canada. *Appl. Environ. Microbiol.* **76**, 1486–1496.

Maheshri, N. and O'Shea, E. K. (2007). Living with noisy genes: how cells function reliably with inherent variability in gene expression. *Annu. Rev. Biophys. Biomol. Struct.* **36**, 413–434.

Maleki-Dizaji, S., Rolfe, M., Fisher, P., and Holcombe, M. (2009). A systematic approach to understanding bacterial responses to oxygen using Taverna and Webservices. In: C. T. Lim & J. C. H. Goh (Eds.), *13th International Conference on Biomedical Engineering*, Berlin Heidelberg: Springer.

Malios, R. R., Ojcius, D. M., and Ardell, D. H. (2009). An iterative strategy combining biophysical criteria and duration hidden Markov models for structural predictions of *Chlamydia trachomatis* σ66 promoters. *BMC Bioinformatics* **10**, 271.

Manning, C. D. and Schutze, H. S. (1999). *Foundations of Statistical Natural Language Processing*. Cambridge, MA: MIT press.

Mazurie, A., Bottani, S., and Vergassola, M. (2005). An evolutionary and functional assessment of regulatory network motifs. *Genome Biol.* **6**, R35.

Mcadams, H. and Arkin, A. P. (1997). Stochastic mechanisms in gene expression. *Proc. Natl. Acad. Sci.* **94**, 814–819.

McCulloch, W. and Pitts, W. (1943). A logical calculus of the ideas immanent in nervous activity. *Bull. Math. Biol.* **5**, 115–133.

Microbiology by numbers [Editorial], (2011). *Nat. Rev. Microbiol.* **9**, 628.

Mostafavi, S., Ray, D., Warde-Farley, D., Grouios, C., and Morris, Q. (2008). GeneMANIA: a real-time multiple association network integration algorithm for predicting gene function. *Genome Biol.* **9**, S4.

Moszer, I., Jones, L., Moreira, S., Fabry, C., and Danchin, A. (2002). Subtilist: the reference database for the *Bacillus subtilis* genome. *Nucleic Acids Res.* **30**, 62–65.

Nabieva, E., Jim, K., Agarwal, A., Chazelle, B., and Singh, M. (2005). Whole-proteome prediction of protein function via graph-theoretic analysis of interaction maps. *Bioinformatics* **21**, 302–310.

Nannapaneni, P., Hertwig, F., Depke, M., Hecker, M., Mäder, U., Völker, U., Steil, L., and Van Hijum, S A F T (2012). Defining the structure of the general stress regulon of *Bacillus subtilis* using targeted microarray analysis and random forest classification. *Microbiology* **158**, 696–707.

Nariai, N., Kolaczyk, E. D., and Kasif, S. (2007). Probabilistic protein function prediction from heterogeneous genome-wide data. *PLoS One* **2**, e337.

Nelofer, R., Ramanan, R., Rahman, R., Basri, M., and Ariff, A. (2012). Comparison of the estimation capabilities of response surface methodology and artificial neural network for the optimization of recombinant lipase production by *E. coli* BL21. *J. Ind. Microbiol. Biotechnol.* **39**, 243–254.

Newman, J. R. S., Ghaemmaghami, S., Ihmels, J., Breslow, D. K., Noble, M., Derisi, J. L., and Weissman, J. S. (2006). Single-cell proteomic analysis of *S. cerevisiae* reveals the architecture of biological noise. *Nature* **441**, 840–846.

Noble, W. S. (2004). Support vector machine applications in computational biology. In *Kernel Methods in Computational Biology*. Cambridge, MA: MIT Press.

Noble, W. S. (2006). What is a support vector machine? *Nat. Biotechol.* **24**, 1565–1567.

Noble, P. A., Bidle, K. D., and Fletcher, M. (1997). Natural microbial community compositions compared by by a back-propagating neural network and cluster analysis of 5SrRNA. *Appl. Environ. Microbiol.* **63**, 1762–1770.

Oinn, T., Addis, M., Ferris, J., Marvin, D., Senger, M., Greenwood, M., Carver, T., Glover, K., Pocock, M. R., Wipat, A., and Li, P. (2004). Taverna: a tool for the composition and enactment of bioinformatics workflows. *Bioinformatics* **20**, 3045–3054.

Outbreak Genomics [Editorial], (2011). *Nat. Biotechol.* **29**, 769.

Ozasa, K., Aono, M., Maeda, M., and Hara, M. (2009). Simulation of neurocomputing based on photophobic reactions of Euglena: toward microbe–based neural network computing. In C. Calude, J. Costa, N. Dershowitz, E. Freire & G. Rozenberg (Eds.), *Unconventional Computation*. Berlin/Heidelberg: Springer.

Palaniappan, K. and Mukherjee, S. (2011). Predicting "essential" genes across microbial genomes: a machine learning approach. In: *10th International Conference on Machine Learning and Applications*.

Paulson, D. S. (2008). *Biostatistics and Microbiology: A Survival Manual*. Springer.

Punta, M., Coggill, P. C., Eberhardt, R. Y., Mistry, J., Tate, J., Boursnell, C., Pang, N., Forslund, K., Ceric, G., Clements, J., Heger, A., Holm, L., Sonnhammer, E. L. L., Eddy, S. R., Bateman, A., and Finn, R. D. (2012). The Pfam protein families database. *Nucleic Acids Res.* **40**, D290–D301.

Qin, J., Li, R., Raes, J., Arumugam, M., Burgdorf, K. S., Manichanh, C., Nielsen, T., Pons, N., Levenez, F., Yamada, T., Mende, D. R., Li, J., Xu, J., Li, S., Li, D., Cao, J., Wang, B., Liang, H., Zheng, H., Xie, Y., Tap, J., Lepage, P., Bertalan, M., Batto, J.-M., Hansen, T., Le Paslier, D., Linneberg, A., Nielsen, H. B., Pelletier, E., Renault, P., Sicheritz-Ponten, T.,

Turner, K., Zhu, H., Yu, C., Li, S., Jian, M., Zhou, Y., Li, Y., Zhang, X., Li, S., Qin, N., Yang, H., Wang, J., Brunak, S., Dore, J., Guarner, F., Kristiansen, K., Pedersen, O., Parkhill, J., Weissenbach, J., Bork, P., Ehrlich, S. D., and Wang, J. (2010). A human gut microbial gene catalogue established by metagenomic sequencing. *Nature* **464**, 59–65.

Quinlan, R. (1993). *C4.5: Programs for Machine Learning*. Morgan Kaufmann.

Rashid, M., Saha, S., and Raghava, G. P. S. (2007). Support vector machine-based method for predicting subcellular localization of mycobacterial proteins using evolutionary information and motifs. *BMC Bioinformatics* **8**, 337.

Rattray, J., Floros, J. D., and Linton, R. H. (1999). Computer-aided microbial identification using decision trees. *Food Control* **10**, 107–116.

Rausch, C., Weber, T., Kohlbacher, O., Wohlleben, W., and Huson, D. H. (2005). Specificity prediction of adenylation domains in nonribosomal peptide synthetases (NRPS) using transductive support vector machines (TSVMs). *Nucleic Acids Res.* **33**, 5799–5808.

Ravcheev, D. A., Best, A. A., Tintle, N., Dejongh, M., Osterman, A. L., Novichkov, P. S., and Rodionov, D. A. (2011). Inference of the transcriptional regulatory network in *Staphylococcus aureus* by integration of experimental and genomics-based evidence. *J. Bacteriol.* **193**, 3228–3240.

Robinson, A. (2007). *The Last Man Who Knew Everything*. Oneworld Publications.

Rohde, H., Qin, J., Cui, Y., Li, D., Loman, N. J., Hentschke, M., Chen, W., Pu, F., Peng, Y., Li, J., Xi, F., Li, S., Li, Y., Zhang, Z., Yang, X., Zhao, M., Wang, P., Guan, Y., Cen, Z., Zhao, X., Christner, M., Kobbe, R., Loos, S., Oh, J., Yang, J., Danchin, A., Gao, G. F., Song, Y., Li, Y., Yang, H., Wang, J., Xu, J., Pallen, M. J., Wang, J., Aepfelbacher, M., and Yang, R. (2011). Open-source genomic analysis of shiga-toxin–producing *E. coli*. *N. Engl. J. Med.* **365**, 718–724.

Rooney-Varga, J. N., Anderson, R. T., Fraga, J. L., Ringelberg, D., and Lovley, D. R. (1999). Microbial communities associated with anaerobic benzene degradation in a petroleum-contaminated aquifer. *Appl. Environ. Microbiol.* **65**, 3056–3063.

Rosales-Colunga, L. M., García, R. G., and De León Rodríguez, A. (2010). Estimation of hydrogen production in genetically modified *E. coli* fermentations using an artificial neural network. *Int. J. Hydr. Energy* **35**, 13186–13192.

Ross, I. L., Browne, C. R., and Hume, D. A. (1994). Transcription of individual genes in eukaryotic cells occurs randomly and infrequently. *Immunol. Cell Biol.* **72**, 177–185.

Sarker, M., Talcott, C., Madrid, P., Chopra, S., Bunin, B. A., Lamichhane, G., Freundlich, J. S., and Ekins, S. (2012). Combining chemoinformatics methods and pathway analysis to identify molecules with whole-cell activity against *Mycobacteriun tuberculosis*. *Pharm. Res.* **29**, 2115–2127.

Schippa, S., Lebba, V., Barbato, M., Di Nardo, G., Totino, V., Checchi, M. P., Longhi, C., Maiella, G., Cucchiara, S., and Conte, M. P. (2010). A distinctive 'microbial signature' in celiac pediatric patients. *BMC Microbiol.* **10**, 175.

Schloss, P. D. and Handelsman, J. (2005). Metagenomics for studying unculturable microorganisms: cutting the Gordian knot. *Genome Biol.* **6**, 229.

Schwikowski, B., Uetz, P., and Fields, S. (2000). A network of interacting proteins in yeast. *Nat. Biotechnol.* **18**, 1257–1261.

Shannon, C. E. (1948). A mathematical theory of communication. *Bell Syst. Techn. J.* **27** (379–423), 623–656.

Shao, Y., Harrison, E. M., Bi, D., Tai, C., He, X., Ou, H.-Y., Rajakumar, K., and Deng, Z. (2011). TADB: a web-based resource for Type 2 toxin–antitoxin loci in bacteria and archaea. *Nucleic Acids Res.* **39**, D605–D611.

Sharan, R., Ulitsky, I., and Shamir, R. (2007). Network-based prediction of protein function. *Mol. Syst. Biol.* **3**, 88.

Shen-Orr, S. S., Milo, R., Mangan, S., and Alon, U. (2002). Network motifs in the transcriptional regulation network of *Escherichia coli*. *Nat. Genet.* **31**, 64–68.

Shuang, D., Yangge, T. and Hao, Z. (2009). Using the bibliometric analysis to evaluate global scientific production of data mining papers. First International Workshop on Database Technology and Applications, 25–26 April 2009. pp. 233–238.

Silva, R., Ferreira, S., Bonifácio, M. J., Dias, J. M. L., Queiroz, J. A., and Passarinha, L. A. (2012). Optimization of fermentation conditions for the production of human soluble catechol-O-methyltransferase by *Escherichia coli* using artificial neural network. *J. Biotechnol.* **160**, 161–168.

Sima, C. and Dougherty, E. R. (2008). The peaking phenomenon in the presence of feature-selection. *Patt. Recogn. Lett.* **29**, 1667–1674.

Singh, J., Kumar, D., Ramakrishnan, N., Singhal, V., Jervis, J., Garst, J. F., Slaughter, S. M., Desantis, A. M., Potts, M., and Helm, R. F. (2005). Transcriptional response of *Saccharomyces cerevisiae* to desiccation and rehydration. *Appl. Environ. Microbiol.* **71**, 8752–8763.

Singh, V., Khan, M., Khan, S., and Tripathi, C. (2009). Optimization of actinomycin V production by *Streptomyces triostinicus* using artificial neural network and genetic algorithm. *Appl. Microbiol. Biotechnol.* **82**, 379–385.

Sintchenko, V., Gallego, B., Chung, G., and Coiera, E. (2009). Towards bioinformatics assisted infectious disease control. *BMC Bioinformatics* **10**, S10.

Sintchenko, V., Anthony, S., Phan, X.-H., Lin, F., and Coiera, E. W. (2010). A Pub-Med wide associational study of infectious diseases. *PLoS One* **5**, e9535.

Sleator, R. D. (2012). Prediction of protein functions. *Methods Mol. Biol.* **815**, 15–24.

Smith, B., Ashburner, M., Rosse, C., Bard, J., Bug, W., Ceusters, W., Goldberg, L. J., Eilbeck, K., Ireland, A., Mungall, C. J., Leontis, N., Rocca-Serra, P., Ruttenberg, A., Sansone, S.-A., Scheuermann, R. H., Shah, N., Whetzel, P. L., and Lewis, S. (2007). The OBO Foundry: coordinated evolution of ontologies to support biomedical data integration. *Nat Biotechnol.* **25**, 1251–1255.

Snir, S. and Tuller, T. (2009). The Net-HMM approach: phylogenetic network inference by combining maximum likelihood and Hidden Markov Models. *J. Bioinform. Comput. Biol.* **7**, 625–644.

Sobhani, I., Tap, J., Roudot-Thoraval, F., Roperch, J. P., Letulle, S., Langella, P., Corthier, G., Van Nhieu, J. T., and Furet, J. P. (2011). Microbial dysbiosis in colorectal cancer (CRC) patients. *PLoS One* **6**, e16393.

Stark, C., Breitkreutz, B.-J., Chatr-Aryamontri, A., Boucher, L., Oughtred, R., Livstone, M. S., Nixon, J., Van Auken, K., Wang, X., Shi, X., Reguly, T., Rust, J. M., Winter, A., Dolinski, K., and Tyers, M. (2011). The BioGRID Interaction Database: 2011 update. *Nucleic Acids Res.* **39**, D698–D704.

Stelling, J. (2004). Mathematical models in microbial systems biology. *Curr. Opin. Microbiol.* **7**, 513–518.

Stevens, R., Goble, C. A., and Bechhofer, S. (2000). Ontology-based knowledge representation for bioinformatics. *Brief. Bioinform.* **1**, 398–414.

Swain, P. S., Elowitz, M. B., and Siggia, E. D. (2002). Intrinsic and extrinsic contributions to stochasticity in gene expression. *Proc. Natl. Acad. Sci.* **99**, 12795–12800.

Szklarczyk, D., Franceschini, A., Kuhn, M., Simonovic, M., Roth, A., Minguez, P., Doerks, T., Stark, M., Muller, J., Bork, P., Jensen, L. J., and Von Mering, C. (2011). The STRING database in 2011: functional interaction networks of proteins, globally integrated and scored. *Nucleic Acids Res.* **39**, D561–D568.

Tettelin, H., Masignani, V., Cieslewicz, M. J., Donati, C., Medini, D., Ward, N. L., Angiuoli, S. V., Crabtree, J., Jones, A. L., Durkin, A. S., Deboy, R. T., Davidsen, T. M., Mora, M., Scarselli, M., Margarit Y Ros, I., Peterson, J. D., Hauser, C. R., Sundaram, J. P., Nelson, W. C., Madupu, R., Brinkac, L. M., Dodson, R. J., Rosovitz, M. J., Sullivan, S. A., Daugherty, S. C., Haft, D. H., Selengut, J., Gwinn, M. L., Zhou, L., Zafar, N., Khouri, H., Radune, D., Dimitrov, G., Watkins, K., O'Connor, K. J. B., Smith, S., Utterback, T. R., White, O., Rubens, C. E., Grandi, G., Madoff, L. C., Kasper, D. L., Telford, J. L., Wessels, M. R., Rappuoli, R., and Fraser, C. M. (2005). Genome analysis of multiple pathogenic isolates of Streptococcus agalactiae: implications for the microbial "pan-genome" *Proc. Natl. Acad. Sci. U.S.A.* **102**, 13950–13955.

Troyanskaya, O. G., Dolinski, K., Owen, A. B., Altman, R. B., and Botstein, D. (2003). A Bayesian framework for combining heterogeneous data sources for gene function prediction (in *Saccharomyces cerevisiae*). *Proc. Natl. Acad. Sci.* **100**, 8348–8353.

Tseng, G. C., Ghosh, D., and Feingold, E. (2012). Comprehensive literature review and statistical considerations for microarray meta-analysis. *Nucleic Acids Res.* **40**, 3785–3799.

Wang, P. I. and Marcotte, E. M. (2010). It's the machine that matters: predicting gene function and phenotype from protein networks. *J. Proteomics* **73**, 2277–2289.

Wang, J., Cheng, D., Zeng, N., Xia, H., Fu, Y., Yan, D., Zhao, Y., and Xiao, X. (2010). Application of microcalorimetry and principal component analysis. *J. Thermal Analy. Calor.* **102**, 137–142.

Warren, L. (1998). *Joseph Leidy: The Last Man Who Knew Everything*. Yale University Press.

Weile, J., James, K., Hallinan, J., Cockell, S. J., Lord, P., Wipat, A., and Wilkinson, D. (2012). Bayesian integration of networks without Gold Standards. *Bioinformatics* **28**, 1495–1500.

Wilkening, J. (2009). Using clouds for metagenomics: a case study. In: *IEEE International Conference on Cluster Computing*.

Witten, I. H., Frank, E., and Hall, M. A. (2011). *Data Mining: Practical Machine Learning Tools and Techniques*. Morgan Kaufmann.

Woese, C. R. and Fox, G. E. (1977). Phylogenetic structure of the prokaryotic domain: the primary kingdoms. *Proc. Natl. Acad. Sci.* **74**, 5088–5090.

Wu, G. D., Chen, J., Hoffmann, C., Bittinger, K., Chen, Y.-Y., Keilbaugh, S. A., Bewtra, M., Knights, D., Walters, W. A., Knight, R., Sinha, R., Gilroy, E., Gupta, K., Baldassano, R., Nessel, L., Li, H., Bushman, F. D., and Lewis, J. D. (2011). Linking long-term dietary patterns with gut microbial enterotypes. *Science* **334**, 105–108.

Wuchty, S., Oltvai, Z. N., and Barabási, A. L. (2003). Evolutionary conservation of motif constituents in the yeast protein interaction network. *Nat. Genet.* **35**, 176–179.

Xiang, Z., Zheng, W., et al. (2006). BBP: *Brucella* genome annotation with literature mining and curation. *BMC Bioinformatics* **7**(347).

Xu, M., Wu, W.-M., Wu, L., He, Z., Van Nostrand, J. D., Deng, Y., Luo, J., Carley, J., Ginder-Vogel, M., Gentry, T. J., Gu, B., Watson, D., Jardine, P. M., Marsh, T. L., Tiedje, J. M., Hazen, T., Criddle, C. S., and Zhou, J. (2010). Responses of microbial community functional structures to pilot-scale uranium *in situ* bioremediation. *ISME J.* **4**, 1060–1070.

Yandell, M. D. and Majoro, W. H. (2002). Genomics and natural language processing. *Nat. Rev. Genet.* **3**, 601–606.

Yooseph, S., Sutton, G., Rusch, D. B., Halpern, A. L., Williamson, S. J., Remington, K., Eisen, J. A., Heidelberg, K. B., Manning, G., Li, W., Jaroszewski, L., Cieplak, P., Miller, C. S., Li, H., Mashiyama, S. T., Joachimiak, M. P., Van Belle, C., Chandonia, J. M., Soergel, D. A., Zhai, Y., Natarajan, K., Lee, S., Raphael, B. J., Bafna, V., Friedman, R., Brenner, S. E., Godzik, A., Eisenberg, D., Dixon, J. E., Taylor, S. S., Strausberg, R. L., Frazier, M., and Venter, J. C. (2007). The Sorcerer II global ocean sampling expedition: expanding the universe of protein families. *PLoS Biol.* **5**, e16.

Yooseph, S., Li, W., and Sutton, G. (2008). Gene identification and protein classification in microbial metagenomic sequence data via incremental clustering. *BMC Bioinformatics* **9**, 182.

You, L., Cox, R. S., Weiss, R., and Arnold, F. H. (2004). Programmed population control by cell-cell communication and regulated killing. *Nature* **428**, 868–871.

Zaremba, S., Ramos-Santacruz, M., Hampton, T., Shetty, P., Fedorko, J., Whitmore, J., Greene, J. M., Perna, N. T., Glasner, J. D., Plunkett, G., Shaker, M., and Pot, D. (2009). Text-mining of PubMed abstracts by natural language processing to create a public knowledge base on molecular mechanisms of bacterial enteropathogens. *BMC Bioinformatics* **10**, 177.

Zhang, T. and Fang, H. H. P. (2000). Digitization of DGGE (denaturing gradient gel electrophoresis) profile and cluster analysis of microbial communities. *Biotechnol. Lett.* **22**, 399–405.

Zhang, C.-B., Wang, J., Liu, W.-L., Zhu, S.-X., Ge, H.-L., Chang, S. X., Chang, J., and Ge, Y. (2010a). Effects of plant diversity on microbial biomass and community metabolic profiles in a full-scale constructed wetland. *Ecol. Eng.* **36**, 62–68.

Zhang, W., Li, F., and Nie, L. (2010b). Integrating multiple 'omics' analysis for microbial biology: application and methodologies. *Microbiology* **156**, 287–301.

Zweigenbaum, P., Demner-Fushman, D., Yu, H., and Cohen, K. B. (2007). Frontiers of biomedical text mining: current progress. *Brief. Bioinform.* **8**, 358–375.

CHAPTER 3

Proteomics: From relative to absolute quantification for systems biology approaches

Andreas Otto*, Jörg Bernhardt*, Michael Hecker*, Uwe Völker[†], Dörte Becher*,[1]

*Ernst-Moritz-Arndt-University Greifswald, Institute for Microbiology, Greifswald, Germany
[†]Ernst-Moritz-Arndt-University Greifswald, Interfaculty Institute for Genetics and Functional Genomics, Greifswald, Germany
[1]Corresponding author. e-mail address: dbecher@uni-greifswald.de

Abbreviations

2DE	two-dimensional gel electrophoresis
LC–MS/MS	liquid chromatography coupled with tandem mass spectrometry
PTM	post-translational modification
SRM	selected reaction monitoring

1 INTRODUCTION

1.1 'Omics' techniques in systems biology approaches

In systems biology, the generation of large-scale interdependent data requires the performance of 'omics' type studies (Ideker *et al.*, 2001). Genome, transcriptome, proteome and metabolome level data are combined to provide a comprehensive set of descriptive data for the biological system under investigation (Kitano, 2002). Even though each individual 'omics' level data reveals unique information and targets different levels of biological regulation, the analysis of the proteome leads to particularly valuable information due to the nature of proteins as the effectors of cellular metabolism and structural determinants throughout all kingdoms of life (Chuang *et al.*, 2010). Modern proteomics approaches can be grouped into two different workflows addressing the analysis of complex proteome data, namely, gel-based and gel-free proteomics techniques.

1.2 Gel-based proteomics

Gel-based studies rely on two-dimensional gel electrophoresis (2DE) for the resolution of complex protein mixtures according to the pI and molecular weight (MW) of the individual proteins in a sample (Neidhardt, 2011). A major asset of gel-based proteomics is the analysis of intact protein species providing direct access to post-translational modification (PTM) events, for example, changes in MW due to proteolytic processing/degradation or direct staining of phosphorylation events by ProQ Diamond (Hecker et al., 2008, 2009; Rabilloud et al., 2009). Differential gel image analysis of 2D gels representing different proteomic snapshots delivers relative quantitative information solely based on the staining intensities of the protein spots found on the gels. Since the invention of 2DE, this technique has matured to become an exceedingly robust and relatively cheap technique in terms of equipment needed (Westermeier and Marouga, 2005; Rabilloud et al., 2010). In microbiology, 2D gel-based proteomics is commonly used for assessment of adaptation responses to nutrient shifts (Bernhardt et al., 1999), antimicrobial agents (Wenzel and Bandow, 2011) or environmental stresses and starvation (Budde et al., 2006). The advantages of 2D gels for the study of microbial physiology, namely, the visualization of the majority of the proteins involved in biosynthetic pathways and the main metabolic routes, have recently been reviewed for Gram-positive bacteria (Völker and Hecker, 2005; Hecker et al., 2008). Elucidating changes in the amounts of key effectors in metabolism and cellular structure provides valuable insights into the main processes of life and is thus essential for systems biology approaches.

1.3 Gel-free proteomics

With the advent of high-resolution and accurate mass spectrometers, facilitating the analysis of complex peptide mixtures by so-called shot-gun proteomics, gel-free studies have gained exceeding significance in the field (Cox and Mann, 2007, 2011; Michalski et al., 2011). In contrast to the analysis of intact proteins in classic (i.e. gel-based) proteomics experiments, gel-free proteomics starts with an enzymatic digestion of a protein sample. With this step, information on the co-occurrence of PTMs on particular protein species or proteolytic processing is largely lost. The resulting complex peptide mix is then subjected to separation by liquid chromatography and analysis by mass spectrometry. Bioinformatic post-processing of the acquired data maps the peptides to the masses determined by liquid chromatography coupled with tandem mass spectrometry (LC–MS/MS) and, finally, to the assembly/grouping of the peptide data to the protein data. Quantitative information is more difficult to obtain for gel-free proteomics experiments unless stable isotopes are incorporated in the samples prior to analysis (Bantscheff et al., 2007). Gel-free proteomics workflows have evolved rapidly in recent years, circumventing gel-based limitations with respect to MW, pI and/or hydrophobicity, thereby setting new standards in the sensitivity, versatility and comprehensiveness of proteomics studies (Cox and Mann, 2007; de Godoy et al., 2008). In microbial proteomics, gel-free techniques have pushed the limits far beyond those associated with studies relying on classical proteomics techniques with respect to the depth covered in relative quantitative studies. In gel-free proteome studies, relative quantitation can be based on a wide array of techniques (Bantscheff et al., 2012). However, due to the ease with which components in growth media can be exchanged, metabolic labelling has set the standards in microbial gel-free proteomics. In addition to metabolic labelling with ^{15}N and the widely applied SILAC methodology, tagging strategies such as iTRAQ have proven their applicability in large-scale proteome studies (Wolff *et al.*, 2006; Dreisbach *et al.*, 2008; Soufi *et al.*, 2010). For Gram-positive bacteria, such as *Staphylococcus aureus*, *Bacillus subtilis* and *Corynebacterium glutamicum*, in-depth proteome studies with almost complete coverage of the main metabolic pathways, as well as subcellular fractions such as the membrane, surface and secreted proteomes have provided new insights in their physiology and pathophysiology (Becher *et al.*, 2009; Otto *et al.*, 2010; Becher *et al.*, 2011; Poetsch *et al.*, 2011). The main limitation of gel-free proteomics technologies is the requirement for expensive and resource-intense mass spectrometry instrumentation.

1.4 Targeted proteomics

In recent years, in parallel with discovery-driven experiments aimed at cataloguing the largest possible number of identified proteins, targeted methods have been developed for monitoring and quantifying selected proteins of interest (Picotti and Aebersold, 2012). Quantitative, discovery-based mass spectrometric experiments require, in general, the identification of the same subset of proteins in each sample to be compared. Relative quantitative information is then derived from the signal intensities of the ion masses of the peptides that were assigned to the target proteins. Despite the advantage of a large number of identified proteins in every sample, this leads to an inevitable excessive redundancy in identification data.

Limitations in quantification have stimulated the development of alternative approaches that rely on existing peptide identification data and fragmentation patterns in target-oriented proteomics experiments (Schmidt *et al.*, 2009; Domon and Aebersold, 2010). In selected reaction monitoring (SRM)-based targeted proteomics, distinct pairs of precursor ion masses and cognate fragmentation masses, derived from existing datasets or large-scale spectral libraries, are used for sensitive and specific determination of these pairs (transitions) in triple quadrupole mass spectrometers (QQQ). QQQ display comparably low resolution and mass accuracy compared to the high-end mass spectrometric equipment that is used in traditional discovery-based experiments. However, QQQ instrumentation is outstanding with respect to its dynamic range (three to four orders of magnitude of linear response; Kirkpatrick *et al.*, 2005) and speed of acquisition (high duty cycle), leading to good ion statistics and sensitivity. Consequently, large numbers of samples may be interrogated for relative abundance of a predetermined number of peptides without the need for time-consuming discovery-based proteomics experiments. Taken together, these features make targeted mass spectrometry the preferred choice for systems biology approaches (Lange *et al.*, 2008).

1.5 Proteomics based on data-independent acquisition

As a consequence of the need to generate ever more comprehensive proteomics datasets for systems biology approaches and recent developments in mass spectrometric instrumentation, new workflows based on data-independent acquisition have emerged. Here, in contrast to other approaches, such as classical proteomics, shot-gun proteomics or targeted proteomics relying on SRM, datasets of the highest complexity are generated in a completely unbiased and untargeted mode of acquisition (Bensimon et al., 2012). Here, conceptually, two approaches for data generation exist: the first, MS^E, is a vendor-specific acquisition mode, whilst the second approach relies on mass spectrometric equipment being capable of fast acquisitions of larger m/z windows for triggering tandem mass spectra.

MS^E is characterized by alternating scan modes: overview scans targeting the intact peptide masses eluting from a reversed-phase column are constantly alternated with collision-induced dissociation scans, fragmenting all precursor ions simultaneously (Bateman et al., 2002). The bioinformatic post-processing of the data from this mode of acquisition is strictly dependent on a highly reproducible LC–MS set-up. This ensures that the retention times are stable enough to facilitate the reassignment of the elution profiles of peptide fragment masses to their respective precursors in the overview scan (Silva et al., 2006a,b).

The second group of acquisition strategies bypasses the decision step of selecting precursor masses suitable for fragmentation, thereby allowing for a bias-free mode of data acquisition (Venable et al., 2004; Panchaud et al., 2009; Geiger et al., 2010; Panchaud et al., 2011; Gillet et al., 2012). Technically, overview scans are followed by collision-induced dissociation scans fragmenting sections of the gas phase of different size and number. Most popular are the methods of precursor acquisition that are independent of ion count (Panchaud et al., 2011) and the more recent development SWATH™ MS (Gillet et al., 2012). Analyses of this type are heavily dependent on the large amount of identification data that has been previously generated by data-dependent mass spectrometry-based proteomics and which is stored in publicly available repositories (Gillet et al., 2012).

In addition to the evident advantage of the data-independent methods to avoid discrimination of certain peptide ion species in terms of abundance or physico-chemical properties, the datasets generated are ready to use for screening, for example, for unusual PTM or peptides which should be covered in these comprehensive datasets both at the MS1 (overview scan) and MS2 (fragment ions) levels.

2 ABSOLUTE QUANTIFICATION WORKFLOWS IN PROTEOMICS
2.1 Fusion protein-based global scale protein quantification

The need for absolute quantification data for systems biology approaches has led to a number of protocols for generation of such data in different organisms and the application of a range of technologies. In microbiology, groundbreaking work has been published on the cellular level quantification of all proteins of *Saccharomyces cerevisiae*. Early approaches to the determination of absolute protein abundances at a proteome-wide scale relied on GFP-tagging or immuno-affinity approaches targeting the specific epitope tags (Ghaemmaghami et al., 2003; Huh et al., 2003).

Whilst these approaches and the underlying experimental principles were, for a long time, regarded as the gold standard, they are currently only used in specific cases and for validation purposes (Brownridge *et al.*, 2011), mostly because of restricted availability of antibodies, the labour intensiveness of the work or, more seriously, the risk of introducing perturbations that might compromise subsequent systems biology approaches.

2.2 Mass spectrometry-based absolute quantification workflows in proteomics

2.2.1 Absolute quantification: Approaches relying on isotope dilution

Overall, mass spectrometry-based absolute quantification strategies are methodologically based on the concept of stable isotope labelling and the isotope dilution (Brun *et al.*, 2007). A defined amount of heavy, isotope-labelled internal standard peptide is added to a sample of interest. Subsequently, using the known concentration of the spiked-in peptide as a reference, it is possible to infer the concentration of the probed sample peptide by comparing the signal intensities of labelled and unlabelled peptides. Three general approaches have been published for using heavy isotope-labelled internal standards, the 'Absolute QUAntification' method (AQUA; Gerber *et al.*, 2003), the 'protein standard for absolute quantification' method (PSAQ; Brun *et al.*, 2007; Dupuis *et al.*, 2008; Jaquinod *et al.*, 2012) and the 'QconCAT' method (Beynon *et al.*, 2005; Pratt *et al.*, 2006).

Historically, the AQUA method published by Gygi and co-workers (Gerber *et al.*, 2003) was the first to propose the use of synthetic heavy peptides as internal calibration standards (calibrants). This method of providing internal standards is the most commonly used for absolute quantification strategies, incorporating synthetic peptides with high-quality standards/high purities that can be obtained from a variety of vendors at relatively moderate costs. Despite these advantages, problems may arise as a result of the tendency of various peptides to stick to surfaces during handling and by other limitations in the synthesis process of the peptides linked to their nature of the primary sequences (Mirzaei *et al.*, 2008).

An alternative approach to providing a heavy isotope-labelled internal standard is the use of intact proteins (Brun *et al.*, 2007; Dupuis *et al.*, 2008; Jaquinod *et al.*, 2012). This involves expression of the target protein in a heterologous host such as *Escherichia coli*. The standard protein is labelled by growing the heterologous host in a heavy isotope-labelled growth medium, and then purifying to homogeneity. After determination of the heavy-labelled standard protein's concentration by amino acid analysis, it is used as an internal standard protein by spiking it into the test samples and using it to determine the absolute abundance of its light (i.e. unlabelled) endogenous counterpart. This workflow is labour-intensive and heavily dependent on the proteins to be determined, as the biochemical properties and possible compatibility issues with the host affect the purification and handling of the heterologous standard protein. Even though this approach provides excellent quantification, the relatively high costs and labour-intensive processing precludes its use for large-scale studies addressing a large number of target proteins.

To address the immense cost and handling difficulties of large-scale AQUA, the QconCAT technique has been developed as an alternative approach (Beynon et al., 2005; Pratt et al., 2006). Here, an artificial gene is constructed allowing for the expression of a heavy labelled synthetic protein that yields peptides upon tryptic digestion that match their endogenous counterparts. The advantage of the QconCAT approach is the concatenation of quantotypic peptides (peptides that are specific for a protein and may be used for its quantitation) for several target proteins in a single artificial protein at relatively moderate costs, allowing for multiplex quantification approaches. Once established and optimized, the production of internal standards is scalable, providing a robust set of standards for large-scale studies. An additional advantage of the QconCAT approach is its particular suitability for the study of protein stoichiometry because labelled quantotypic peptides derived from the QconCAT protein are present in equimolar amounts.

For both the AQUA and QconCAT approaches, it is crucial to choose appropriate peptides to serve as internal standards. However, the choice of such peptides can be difficult, often requiring iterative optimization processes and integration of multiple levels of information to yield the peptide species that are most appropriate for absolute quantitation studies (Mallick et al., 2007).

2.2.2 Absolute quantification: Approaches relying on label-free quantitation with best flyer peptides

As alternatives to methods based on the isotope dilution concept, several label-free quantification methods have been developed for the generation of absolute quantitative data (Bantscheff et al., 2012). Instead of spiking-in isotope-labelled standards into the sample, another characteristic of proteome samples is exploited. It has been shown that the abundance of a protein in a sample is correlated directly to its signal intensity in mass spectrometry. This property is used in absolute quantification methods like label-free quantification with 'best flyer' (Top3; Silva et al., 2005; Silva et al., 2006a,b) and intensity based absolute quantification (IBAQ; Schwanhäusser et al., 2011) approaches.

In the original paper by Silva et al. (2006a,b), the strong relationship between the concentration of a protein and the intensity of the mass spectrometric signal response for the three most intense tryptic peptides (Hi3) could be used for the absolute quantification of all of the proteins in a sample (Silva et al., 2006a,b). This workflow provides the option of determining absolute protein abundances in relation to a spiked-in reference protein that calibrates a universal signal response factor applied to all proteins. The Hi3 approach can also be used with other data-dependent mass spectrometric acquisition methods that provide sufficient reproducibility with respect to peptide sequencing (Grossmann et al., 2010). More recently, IBAQ was introduced as a method that takes into consideration the ratio between the number of theoretically observable peptides versus the number of actually measured peptides (Schwanhäusser et al., 2011).

2.3 Large-scale absolute quantification in proteomics

Isotope dilution-based absolute quantification strategies are perfectly suited for the determination of small (PSAQ, AQUA) to medium (QconCAT) scale protein quantities, for example, of enzymes of specific biochemical pathways. Although Brownridge *et al.* (2011) have applied the QconCAT method on a large scale in yeast, scaling up of these methods is, in most cases, restricted by the costs of the peptides required if hundreds or even thousands of proteins are targeted for analysis or simply by the time that has to be spent on QconCAT design, the optimization needed for each specific peptide assay and the time needed for data acquisition. Therefore, a number of studies have strived to combine large-scale proteomics data with absolute protein quantification with the aim of yielding absolute proteome data on a large scale (Malmström *et al.*, 2009; Maass *et al.*, 2011; Maier *et al.*, 2011; Schmidt *et al.*, 2011). To achieve this, large-scale proteome datasets are first acquired either by mass spectrometry-based or by 2D gel-based proteomics. Subsequently, relative quantification is carried out for all samples in the study. This could be based on staining intensities in 2D gel-based approaches or on integrated peak signals in mass spectrometry-based approaches. For calibration of a large-scale dataset, absolute amounts of a small subset of proteins, called anchor proteins, are determined by targeted proteome analysis using either spiked-in heavy labelled peptides (AQUA) or peptides generated by QconCAT. The absolute quantification data from the anchor proteins are then propagated to all of the other measured proteins.

The label-free methods relying on the 'best flyer' concept are, *per se*, putative large-scale methods. Here, the proteome coverage is restricted by the capacity of the LC–MS system used for the analysis. In view of the relatively easy implementation and the achievable high-throughput, label-free methods appear to be more attractive for large-scale studies. However, it should be noted that these methods are still less precise than methods based on stable isotope dilution (Bantscheff *et al.*, 2012).

3 THE PROTEOMICS WORKFLOW

3.1 Sample preparation for absolute quantification strategies in proteomics

3.1.1 General considerations

Despite major methodological differences in absolute quantification strategies, the methods for preparing cell extracts are comparable, with only minor changes applied for specific workflows (Figure 3.1A). In general, to fulfil the requirements of different 'omics' techniques, the complete design in the workflow, including, for example, sample collection, storage and processing, must be coordinated throughout the whole experiment. Technical issues may be encountered when combining transcriptome, proteome and metabolome sampling in parallel. Whilst sampling for transcriptome studies requires care to avoid RNA degradation (e.g. inclusion of inhibitors and rapid cooling/killing of cells by harvesting on ice-cold buffer), proteome samples are

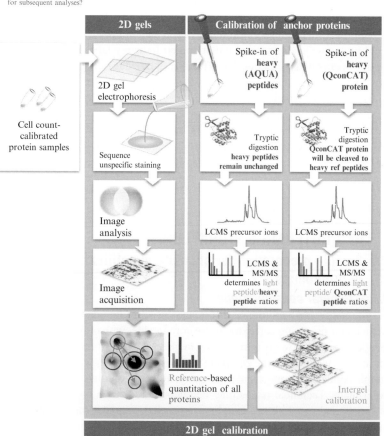

treated with care with respect to their low sample volumes, the need for fast cooling and the rapid removal of residual buffers/media originating from the growth medium. Furthermore, the determination of low MW intracellular and extracellular compounds, as analyzed in metabolome studies, requires the rapid stabilization of their physiological status because metabolite pools tend to be highly dynamic. Additionally, compounds often found in defined growth media, which are known to cause quenching effects, can affect metabolome studies. Consequently, the experimental design starts as early as the choice for media for cultivation and becomes even more important at the sampling and subsequent processing stages.

3.1.2 Cultivation

Systems biology approaches depend on robust protocols for bacterial cell growth. To achieve this, appropriate cell growth media and cultivation conditions have to be tested for robustness and reproducibility. Key cornerstones are the standardized preparation of stock cultures and the use of standard operation procedures for the pre- and main cultures, ensuring little variation in the growth rate as a measure of reproducibility.

3.1.3 Determination of cell counts

Absolute proteomic studies require the number of cells in a sample to be accurately determined at each sampling time. Various methods are available including determining the number of colony-forming units and measuring optical density using a spectrophotometer. In our hands, the determination of cell numbers in a counting chamber has proved to be the gold standard. Unlike methods that rely solely on the density of a suspension, visual counting allows live and dead cells to be distinguished, as well as the detection of unusual cell morphologies. Photographing a series of counting chamber grids helps to shorten the time required for the analysis and helps keeping the cells in the physiological state of interest.

3.1.4 Determination of cell dimensions

The estimation of cell dimensions (e.g. average cell length, cell diameter and cell wall/membrane thickness) is essential for calculations based on the cell volume. These values are determinants of cell morphology and are dependent on the

---

FIGURE 3.1

(A) Preparation of cell extracts for absolute quantification strategies in proteomics. The key features/steps of the preparation of cell extracts for absolute quantitative proteomics for use in systems biology approaches. (B) Large-scale absolute quantitative proteomics with SRM-calibrated 2D gels. Outline of the complete workflow of SRM-calibrated 2DE. The left-hand column shows the workflow for 2DE part of the analysis, leading to the determination of anchor proteins. The workflow leading to the large-scale absolute quantification of the anchor proteins by targeted analysis based on AQUA (middle column) or on QconCAT (right-hand column). (For colour version of this figure, the reader is referred to the online version of this chapter.)

physiological factors such as growth rate and culture conditions. Micrographs can be automatically analyzed using imaging software such as ImageJ (Abramoff et al., 2004) and the distribution of cell length and diameter determined automatically by using the Feret's diameter property of all selected image features (i.e. bacterial cells). The most precise measurement of cell volume is likely to be accomplished using membrane stains that facilitate the calculation of the true intracellular volume.

3.1.5 Cell disruption

Cell disruption is crucial for the release of intracellular components such as nucleic acids, metabolites or proteins. The cell disruption method that is used must be effective in releasing these cellular compounds reliably, efficiently and effectively. In the context of systems biology, special attention has to be paid to disrupting the cells in a very short-time frame, otherwise, the analyses will be biased by effects associated with the analysis workflow rather than the physiology of the cells during cultivation. Additionally, because of the labile nature and the dynamics of biological macromolecules, the sample temperature should be kept to a minimum, ensuring cooling to 4 °C or below during the entire cell disruption process.

A detailed analysis of bacterial cell disruption techniques reveals a clear correlation between disruption efficiency, the species being investigated (including cell wall type) and the physiological state of the cells. Rod-shaped bacteria are easier to disrupt than coccal-shaped cells that form a vault-like structure that is much more resistant to mechanical forces. The cell walls of Gram-positive bacteria are more rigid than those of Gram-negative bacteria, whilst wall-less bacteria (e.g. *Mycoplasma*) can usually be disrupted by hypo-osmotic shock. Stabilizing or surrounding layers such as surface (S) layer proteins and polysaccharide capsules may influence cell disruption efficiency. Moreover, it is known that the Gram-positive bacteria *B. subtilis* and *S. aureus* thicken their cell walls upon entry into stationary phase, resulting in reduced disruption efficiencies compared to exponentially growing cells (Middelberg, 1995). Also, younger (i.e. newly born) cells may be disrupted more easily than older ones. Finally, the disruption process is tightly connected to the system under investigation and needs to be optimized for any species, cell type and cell state. As shown in Figure 3.2, we have tested a variety of cell lysis methods for both *B. subtilis* and *S. aureus* in order to identify the most effective workflow for both exponentially growing and stationary phase cells. By using the optimized parameters, disruption efficiencies $\geq 90\%$ are achievable. However, it is becoming increasingly clear that cell disruption technology still needs to undergo substantial improvements to ensure efficiency and lack of discrimination for cells of differing sizes, shapes and composition.

To ensure efficient and complete cell lysis of each specific cell type, it is advisable to evaluate all the methods available in the respective laboratories. Benchmarking cell disruption efficiencies may be based either on the determination of colony-forming units or on the counting of intact cells in a counting chamber. Furthermore, cell viability assay based on, for example, fluorescence markers may also be used.

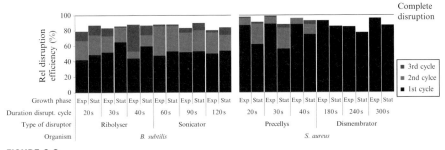

FIGURE 3.2

Evaluation of disruption efficiencies. Different cell lysis methods for the Gram-positive bacteria *B. subtilis* and *S. aureus* have been tested for both exponentially growing cells and stationary phase cells. By using optimized lysis parameters, disruption efficiencies of $\geq 90\%$ are achievable. (For colour version of this figure, the reader is referred to the online version of this chapter.)

3.1.6 Determination of protein concentration

The determination of protein concentration is crucial in absolute quantitative workflows. The failure to consider potential variations would lead to errors in precision and accuracy that would be amplified in all downstream calculations aimed at determining protein content.

Prior to any proteome analysis, it is strongly recommended to undertake an evaluation of the existing methods for determining protein content. A vast number of commercial systems exist that are based on the physical interactions or biochemical properties of proteins. Although the choice for a specific system is likely to be dependent on existing and widely available methods, consideration still needs to be given to appropriateness of the chosen assay. Such considerations include the presence of interfering chemicals and components in the buffers/media used, as well as the average amino acid composition of the proteins under investigation. As shown in Figure 3.3, we have compared a commercially available Bradford assay with an assay that is based on the Ninhydrin reagent. The commonly used Bradford assay is cheap, easy to use and sensitive. The active ingredient is Coomassie Brilliant Blue G250, a disulfonylated triphenylmethane dye, that binds to proteins at positively charged amino acids (arginine, lysine and histidine) via its sulfonyl groups and at hydrophobic amino acids (tryptophan, tyrosine and phenylalanine) and aliphatic chains of SDS within the protein via its triphenylmethane backbone (Tal *et al.*, 1985). This makes it sensitive to sequence variations, especially for amino acids that are involved in the electrostatic and hydrophobic binding of the dye. The traditional protein standard bovine serum albumin should not be used in connection with Coomassie G250 because it has a much higher affinity for this dye compared to other proteins.

The issue of biased protein content with respect to quantification can be circumvented by using detection agents that do not show specificity for sequence variations.

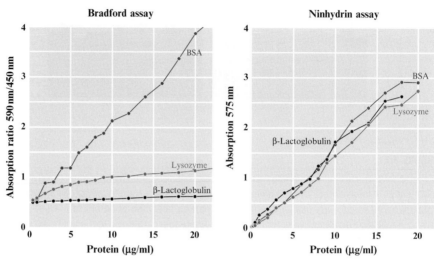

FIGURE 3.3

Evaluation of different assays for the determination of protein concentration. Absorption curves have been determined for different standard proteins (BSA, bovine serum albumin; β-LG, beta-lactoglobulin; CSA, chicken serum albumin) with two different protein concentration assays. These clearly show that the Ninhydrin-based protein assay outperforms the Bradford assay in terms of linearity and lacking bias towards specific proteins. (For colour version of this figure, the reader is referred to the online version of this chapter.)

This means that, specific protein features aside, the primary sequence alone should be used for the precise determination of protein concentration. The peptide bond exactly fits this need. By using acid hydrolysis, the alpha amino groups can be released from the peptide bonds and used as the reactive group to form the purple coloured imino derivative of Ninhydrin. Photometric measurement of the purple colour gives a proportional value of the amount of peptide bonds and, thereby, of the amount of protein within the sample (Figure 3.3).

4 GENERATION OF ABSOLUTE QUANTITATIVE DATA BY TARGETED MASS SPECTROMETRY

4.1 Criteria for selection of peptides for targeted analyses

One of the approaches to determining the absolute protein abundance in a proteome sample is targeted analysis by SRM, relying on the stable isotope dilution concept. The key practical consideration that needs to be considered for such analyses is which proteins to target. This requires gathering as much information as possible from previous proteomic experiments about potential targets. Because the peptides used to represent the target proteins serve as surrogates for quantitation (Kirkpatrick et al., 2005), a number of additional factors need to be considered to ensure that other relevant criteria are fulfilled. These include the general requirements of targeted mass spectrometry analyses, criteria specific to studies based on synthetic peptides (AQUA) or criteria specific to the QconCAT approach. These factors are considered below.

Targeted mass spectrometry relies on the high specificity of precursor/fragment ion pairs (transitions) generated in MS/MS experiments in triple quadrupole-based instruments (Lange *et al.*, 2008). Adding to the mass filtering, these transitions are measured in predetermined time windows during LC–MS runs to allow for multiplexed SRM analyses. To facilitate scheduled SRM assays, it is essential to know the elution times of the peptides chosen for targeted analyses. Consequently, if not already determined from previous studies, the column retention time of each candidate peptide that is selected for SRM analysis has to be determined for the various chromatography formats. With the two levels of mass selection (i.e. Q1/Q3; precursor/fragment ion) and the appropriate time window established, it is possible to achieve the required high specificity of co-eluting masses with very high sensitivity and low background.

The technically challenging aspect of targeted proteomics workflows is to set up validated SRM assays for peptides representing the specific proteins of interest (Kettenbach *et al.*, 2011). Fragmentation patterns allowing selection of appropriate peptides for a protein of interest are available either in public repositories (Desiere *et al.*, 2005; Martens *et al.*, 2005) or in existing MS/MS datasets. These sources yield information on those peptides that are, in practice, good responders in an LC–MS/MS experiment. For peptides that have been identified previously in such experiments, it is important to recognize that larger peptides leading to high confidence scores are not necessarily good candidates for SRM pairs. Here, the selection of smaller peptides and peptides containing proline residues (incomplete fragmentation patterns) will produce higher intensity fragment masses and thereby provide better transitions. Problems may arise for peptides whose sequences are shared with homologous proteins or those sharing the same domains. Therefore, the chosen peptides must be proteotypic, that is, they must uniquely identify the targeted protein (Mallick *et al.*, 2007).

In addition to their unique identity, it needs to be recognized that peptides may be identified on the basis of more than just their primary sequence. Peptides may be post-translationally modified, for example, by the oxidation of methionine during sample processing or even specifically in response to a native biological process. Such modified peptides have to be handled with care and, consequently, are mostly excluded from targeted analyses. This criterion is summed up by the term of a 'quantotypic peptide': peptides have to be chosen that are not only proteotypic but also representative for the protein and its changes in abundance throughout the entire analysis (Brownridge *et al.*, 2011).

Most studies that describe the workflow for targeted analyses are dependent on trypsin for proteolytic digestion since, under optimum conditions, complete proteolysis of a sample can be expected (Barnidge *et al.*, 2004; Havlis and Shevchenko, 2004; Olsen *et al.*, 2004). Nevertheless, digestion efficiencies need to be checked

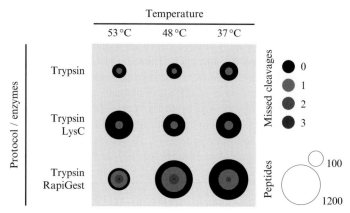

FIGURE 3.4

Evaluation of digestion efficiencies. For the bias-free determination of absolute protein quantities, a complete digestion must be ensured. Detailed testing and optimization of digest conditions, even with trypsin, are advisable. Here, the optimum temperature of 37 °C and the use of the surfactant RapiGest led to the best results for tryptic digests that are suitable for SRM analyses in terms of the highest number of peptides without missed cleavages. (For colour version of this figure, the reader is referred to the online version of this chapter.)

prior to generating quantitative data, as is shown in Figure 3.4. Peptides that result from the incomplete digestion with one or more missed cleavage sites (Brownridge and Beynon, 2011), and peptides with ragged ends (two or more arginine and/or lysine residues in a row), should be avoided as candidate peptides for SRM (Barnidge et al., 2004).

The above criteria will lead to the production of a list of peptides that are both proteotypic for the proteins under investigation and amenable, at least in theory, for SRM analyses. The entire process for the selection of proteins with suitable peptides, proteotypic peptides and transitions, as well as their validation and optimization, will, in most cases, be aided by dedicated software suites (Lange et al., 2008).

In targeted proteomics, absolute quantification is based on the comparison of transition intensities for endogenous peptides versus their heavy-labelled spiked-in counterparts. This may be achieved by spiking-in to the sample artificially synthesized peptides (AQUA) or a QconCAT protein, prior to tryptic digestion to yield appropriate heavy-labelled peptides. Depending on the method for the supply of the isotopically labelled peptides, further restrictions may apply to the choice of peptides needed for the targeted analyses, as discussed below.

4.2 Basic requirements for AQUA

Once a number of proteotypic peptides have been derived and the AQUA method chosen for their production, the focus changes towards feasibility of their chemical synthesis as heavy-labelled peptides. Currently, synthetic peptides of high purity are available from a variety of vendors. In addition to ordering expensive high-quality peptides, it is also possible to order cheaper, low-quality peptides that are suitable for optimizing the SRM assays (Picotti *et al.*, 2009). Aspects to consider when ordering the peptides are the technical feasibility of chemical peptide synthesis, its length (optimum length 6–13 residues) and desired functionality (heavy amino acids incorporated, PTMs) (Mayya *et al.*, 2006). Additionally, the peptides should be completely soluble in aqueous solutions and therefore limited in size and the avoidance of excess hydrophobic residues. To ensure that the peptides remain chemically stable during the analysis under the processing conditions, amino acids that are prone to oxidation, particularly methionine and cysteine, should also be avoided. Furthermore, consideration should be given to the positioning of the isotopically labelled amino acid since, to reduce the noise levels of the signals, it is advantageous if both the precursor and fragment masses of the respective transitions are heavier than their natural counterparts (Kirkpatrick *et al.*, 2005). To minimize issues associated with incomplete resuspension, chemically synthesized peptides should be already dissolved in an appropriate aqueous buffer system when obtained from the vendor. This allows the absolute concentrations in solution to be determined by amino acid analysis.

4.2.1 De novo *gene design for QconCAT*

Relying on the same basic principles for relative and absolute quantification with a spiked-in heavy-labelled standard, the difference between absolute quantification based on AQUA and QconCAT is the source of the peptide standards. Whilst AQUA peptides are the product of chemical synthesis, the QconCAT peptides are proteolytic products derived from a protein that has been synthesized biologically from an artificial gene in a bacterial expression host.

In the design phase for such a QconCAT gene, the Qpeptides representing the proteins under examination (i.e. peptides that are proteotypic) are selected without the need to consider the restrictions that limit the choice of chemically synthesized peptides. Selection criteria relate solely to the uniqueness of the protein in terms of proteotypic peptides, propensity to ionize and detectability in a mass spectrometer. Nevertheless, limitations that arise as a result of the experimental design, such as the presence of the amino acids that can be labelled and the avoidance of primary sequences containing cysteine or methionine that are prone to PTM, still have to be considered. In order to optimize the synthesis of the QconCAT protein, the randomly concatenated *in silico* DNA sequences have to be codon-optimized for expression in the heterologous expression host, usually *E. coli*. Here, a common limitation is the formation of unwanted secondary structures in the RNA transcript that might diminish expression. During the design of the synthetic gene, not only is the sequence of the Qpeptide gene itself considered, but also additional features such as the inclusion of a start codon, a purification tag (e.g. hexahistidine), protected sacrificial regions (to avoid exoprotease activity) and quality control peptides. These additional features allow for affinity enrichment of the expressed protein, and, in particular, the quality control features add multiple layers of control in the subsequent quantification steps to facilitate the precise determination of protein concentration, for example, by comparison against a universal peptide standard or quantification by immuno-affinity methods that target specific peptides found on the QconCAT protein. An excellent protocol that includes the considerations necessary for the design of a QconCAT gene was published by Pratt *et al.*, (2006).

4.2.2 Expression, purification and usage of the QconCAT protein

Following the design phase of the synthetic QconCAT gene, the cognate protein is expressed and, following affinity purification and the determination of its concentration, is ready to use. The latter two steps are crucial since a high degree of purity and the precise determination of concentration are necessary for the optimal quantification of the proteins of interest. Furthermore, as the Qpeptides are generated in equimolar proportions by proteolytic cleavage (at least in theory), a thorough revision of the proteolytic digestion efficiency of the QconCAT protein and the proteome sample must be determined. This will ensure a bias-free analysis without an underestimation of the proteins of interest. This is particularly important since when the Qpeptides are randomly concatenated in the synthetic protein, the natural context of the primary amino acid sequence is abolished. As a result, it is commonly assumed that QconCAT proteins do not form specific secondary structures leaving the proteolytic cleavage sites completely accessible. Consequently, the complete proteolytic cleavage of QconCAT proteins is generally more efficient than that of their natural counterparts. This issue needs to be addressed by adopting specific digestion protocols that completely denature the native and QconCAT proteins and this issue should ideally be addressed in every new QconCAT-based study (Pratt *et al.*, 2006; Rivers *et al.*, 2007; Mirzaei *et al.*, 2008).

4.3 Quantitative SRM assays based on heavy-labelled peptides

With a source of heavy-labelled peptides finally to hand, it is relatively straightforward to check the peptides for their fragmentation patterns, retention time and instrument-specific SRM assay optimization. The parameters that need to be optimized include the choice of best fragments, determination of the prevalent charge state of the precursor, optimization of collision energies and other, vendor-/instrument-specific parameters. Rather than describing the SRM optimization in detail, readers are referred to excellent existing protocols that provide technical guidelines for setting up new SRM assays (Lange *et al.*, 2008; Gallien *et al.*, 2011; Kettenbach *et al.*, 2011).

Regardless of the origin of the heavy-labelled internal standard peptides for SRM, optimization has to be completed prior to starting the actual analysis. In quantitative SRM assays, the optimized and scheduled transitions for the peptides under examination are combined with the methods associated with the mass spectrometric instrument to ensure that all transitions are determined with the maximum number of data points (at least 8–10), minimizing the number of repeated LC–MS runs. In order to derive quantitative data from the signal intensities of the endogenous peptide compared to the spiked-in heavy-labelled peptide, calibration curves have to be acquired using the optimized parameters in order to determine the limits of detection and quantification (Armbruster *et al.*, 1994; Zorn *et al.*, 1997). Once the calibration curves are established, they can be used to determine the peptide-specific linear range of the transition signal versus the peptide concentration of a sample. Additionally, heavy-labelled peptides should be introduced to the protein sample at levels close to the expected natural levels to ensure that the sample concentration falls within range of the assays linear response (Gallien *et al.*, 2011). Whereas this is easily achieved for AQUA, a trade-off is necessary for the QconCAT-derived peptides that are inherently equimolar and therefore their concentration should be adjusted so as to average out over the expected concentration range of the sample peptides. This drawback can be limited if the expected protein concentrations are known prior to the beginning of the analyses, because then proteins of a similar concentration range can be grouped into the same QconCAT protein.

For the quantitative work, a known amount of the heavy peptides or QconCAT is spiked into the proteome sample prior to proteolytic digestion. It is important that a defined amount of protein extract is used if the same sample has previously been subjected to differential gel image analysis. For the targeted measurement of the light and heavy peptides with the same optimized SRM assays, the areas under the curves for the two counterparts are used for the determination of the absolute expression level of the peptide and therefore of the concentration of the protein in the sample (Kirkpatrick *et al.*, 2005; Gallien *et al.*, 2011; Kettenbach *et al.*, 2011).

The overall goal of setting up the targeted analysis is to determine the best transition pairs, with respect to the optimized instrument parameters and specificity in the chosen retention time window, over other ion species that may be found also in a targeted SRM approach. The quantitative data derived from experiments based on AQUA peptides or on the QconCAT approach give information about the concentration of a given peptide in the interrogated sample, which can also serve as the basis for calculating the absolute abundance of related proteins in the proteome sample (see below).

5 GENERATION OF LARGE-SCALE RELATIVE PROTEOMICS DATA: DIFFERENTIAL 2D GEL IMAGE ANALYSIS

Differential 2D gel image analysis is a widely used method for generating large-scale relative quantitative proteomics abundance data. Despite being the oldest proteomics method, 2DE is still a workhorse for laboratories that rely on its distinct advantages (Rabilloud *et al.*, 2010).

2DE of proteins is technically characterized by a continuous separation space within predefined *p*I- and MW ranges, performed by an electrophoretic step during isoelectric focusing (IEF) and subsequent resolution by MW by gel electrophoresis (Carrette *et al.*, 2006). The main advantages of 2DE, compared to other protein separation approaches, are the direct accessibility and visualization of protein modifications without the need to predict them, the reasonable financial and time costs, and its wide usage in a large number of laboratories. In general, differential 2D gel image analysis is highly dependent on the use of reproducible experimental conditions. Reproducibility is important not only for the consistent separation of proteins in both dimensions of the 2DE but also to avoid analytical bias that can result from technical issues, including rehydration of the IPG-strip, IEF and transfer from the first to the second dimension.

A critical point in 2D gel analysis is the choice of the dyes used to stain the separated proteins. For the quantitative in-gel determination of protein amounts, it is important to be able to predict the behaviour of the dyes used, particularly with respect to the response curve that characterizes the relationship between the protein amount and the detectable signal. Ideally, the dye signal should be independent of the protein sequence and linear over a wide range of protein concentrations. As indicated in Figure 3.5, we have shown that Krypton and Flamingo, two highly sensitive fluorescent dyes, meet the requirements of a lack of sequence bias, excellent sensitivity and wide dynamic range. Other frequently used dyes such as Coomassie, or staining techniques using silver nitrate, do not fully meet these criteria and are therefore not recommended.

Having resolved the proteome samples on 2D gels, state-of-the-art image analysis techniques are used for differential 2D gel image analysis (Berth *et al.*, 2007). Here,

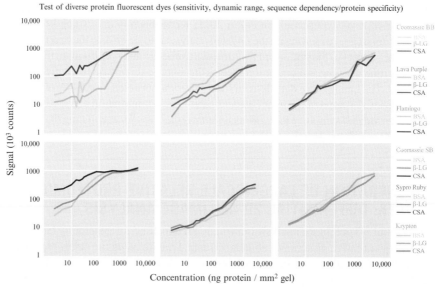

FIGURE 3.5

Test of diverse fluorescent dyes. Signal intensities over a range of concentrations are plotted for six different dyes commonly used in 2DE. As can be seen, the sensitivity and dynamic range is dependent on both the specific dyes and the protein species used for the assay (BSA, bovine serum albumin; β-LG, beta-lactoglobulin; CSA, chicken serum albumin). (For colour version of this figure, the reader is referred to the online version of this chapter.)

sophisticated software suites are used to extract 'real' protein signals and to discriminate them from artificial signals such as image noise, fluorescent speckles and other disrupting features. Following the detection of genuine signals, the images are processed to generate consensus spot patterns of all the images within the experiment (Luhn et al., 2003). For this purpose, the signal intensities of all gel images are combined to generate an artificial proteome map from which the spot boundaries are determined. Matrix assisted laser desorption/ionisation based mass spectrometry is then used to identify the proteins in as many of the detected spots of the proteome map as possible. With these spot identifications at hand, complemented by the established spot boundaries of the proteome map and a series of positionally corrected 2D gel images, it is possible to extract comprehensive, relative expression profiles.

6 LARGE-SCALE ABSOLUTE QUANTITATIVE PROTEOMICS WITH SRM-CALIBRATED 2D PAGE

As introduced above, several approaches have been developed for large-scale absolute quantification of proteomic data. All of these approaches are suitable for the quantification of the soluble proteins mostly found in the cytosolic cell extracts, effectively guaranteeing a reproducible proteome analysis for this cellular subfraction of proteins. Although in part these methods generate their data in markedly differing ways, theoretically, the results should be comparable. Some of the methods rely heavily on high-end mass spectrometric equipment, whilst others use relatively basic and widely available equipment. Consequently, there is a trade-off between methods that provide very precise quantitative data, but at a high cost, and those in which the same level of precision cannot be achieved, but involve considerably less effort and resources. Here, we consider a combination of the classical proteomics approach to 2D differential image analysis as a well-established and relatively cost–effective method for analysing complex proteome samples, and a targeted proteomics approach based on SRM and relatively basic mass spectrometric equipment. Having already introduced the methodologies for SRM analyses based on AQUA and QconCAT, as well as differential 2D gel image analysis, we now focus on the steps necessary to integrate these methods to quantify soluble protein separated by 2DE.

In a study relying on SRM-calibrated 2DE gels, all quantitative aspects of sample preparation have to be tracked, including cell count, cell size/volume, efficiency of cell disruption, determination of protein content and compensation for protein loss at different preparation steps by using spiked-in protein standards from the beginning. All these steps are summarized in Figure 3.1A. With this extensively characterized protein sample at hand, two branches of the proteome workflow have to be followed, as shown in Figure 3.1B.

In the first place, a 2DE analysis is carried out for all proteome samples under investigation. As a result, relative protein expression profiles are generated that can be used subsequently for calibration with the targeted proteomics part of the workflow. Working under the assumption that all proteins behave similarly with respect to their entry into the gel and their binding of fluorescent dye, the use of calibration proteins which are spiked into the samples before 2D separation facilitate the quantitative detection of all of the resolved proteins. Normally, the spot volumes of such calibration proteins would be calculated and expressed as a proportion of total protein extract, as determined from the sum of the individually resolved protein spots. This relative quantitative estimation can then be used to calculate protein amounts independently of the fractional losses of protein extracts during the 2DE process.

In the SRM-calibrated 2D gels, another path is followed: a subset of proteins is chosen from the differential 2D gel image to serve as 'anchor proteins'. These anchor proteins serve as the internal standards in the complete analysis: instead of spiking-in known amounts into the sample as indicated above, the absolute concentration in the sample is determined by targeted SRM analyses. To be a candidate anchor protein, it has to be present on all the 2D gels examined. This is especially important in proteome studies that examine the influence of certain triggers, such as antibiotic stresses, that lead to marked rearrangements of the cell's protein inventory. In this case, a characteristic of a proteome study, namely, the relative stability of the proteomic fraction compared to the transcriptome, represents an advantage. Whilst a number of proteins undergo rapid degradation by a combination of proteases and the cell's quality control system, even under starvation conditions, a significant proportion of proteins remain relatively stable in a time frame that can be measured in hours (Becher *et al.*, 2009; Otto *et al.*, 2010; Michalik *et al.*, 2012). Consequently, proteins that fulfil this criterion will be available in most microbial proteomic studies.

Putative anchor proteins have to form good quality spots that are well separated from other spots on the gels. Sufficient resolution will lead to unambiguous and reliable detection by commercial software packages, complemented either by direct mass spectrometric identification or by spot matching with a pre-assembled master gel/proteome map. To ensure bias-free calibration, the selected anchor proteins should cover the entire range of molecular masses and isoelectric points to be examined on the 2D gels. These criteria are possible in most cases. Proteins that reside in multiple spots or that show a high degree of variability are difficult to quantify reliably and should not be selected as anchor proteins. Such proteins are likely to have undergone PTM or to be targets of protease activity.

After relative quantification of all proteins covered by the 2DE analysis, and, in particular, the anchor proteins, the absolute protein concentrations of the anchor proteins are determined by SRM as previously described (Figure 3.1B). Combining the absolute data from the SRM analysis with the relative data from the differential 2D gel image analysis is possible using two approaches. In the first approach, each proteome sample, and therefore each single gel, is calibrated by itself. This is referred to as intragel calibration. In the second approach, the whole experimental set-up allows for the calibration of only a part of the proteome samples (e.g. a single time point within a complete time series or a control sample in a dose-dependent study) referred to as intergel calibration. Both approaches are equally suited in SRM-calibrated 2DE, as has been shown by Maass *et al.* (2011) (Figure 3.4).

Although the method of SRM-calibrated 2DE is already providing comprehensive and generally precise data, there is still room for improvement. For example, the current SRM-calibrated 2DE workflow could potentially be improved by using difference gel electrophoresis (Ünlü *et al.*, 1997) to reduce the experimental error introduced by the differential gel image analysis. Here, protein samples are stained with different Cy-dyes that allow for the concurrent separation of up to three protein samples on the same 2D gel. This would allow one of the fluorescence channels to be used as a universal proteome reference for the determination of the absolute abundance of the anchor proteins and the other two channels for the proteome samples under investigation.

For systems biology approaches, any information regarding the absolute abundance of proteins in a sample is interesting but not sufficient. The last step in a systems biology proteomics approach will therefore consist of a combination of information from the sample processing step and large-scale absolute quantification step. This allows for the cell volume to be used to determine the amount of protein yielded per cell and the application of correction factors that take into account losses during cell disruption. Finally, the amount of each protein within a single cell, and therefore the copy number per cell of each protein detected on the 2D gel, can be calculated.

References

Abramoff, M., Magalhães, P. J., and Ram, S. (2004). Image processing with ImageJ. *Biophoton. Int.* **11**(7), 36–42.

Armbruster, D. A., Tillman, M. D., and Hubbs, L. M. (1994). Limit of detection (LOD)/limit of quantitation (LOQ): comparison of the empirical and the statistical methods exemplified with GC-MS assays of abused drugs. *Clin. Chem.* **40**(7 Pt. 1), 1233–1238.

Bantscheff, M., Schirle, M., Sweetman, G., Rick, J., and Küster, B. (2007). Quantitative mass spectrometry in proteomics: a critical review. *Anal. Bioanal. Chem.* **389**(4), 1017–1031.

Bantscheff, M., Lemeer, S., Savitski, M. M., and Küster, B. (2012). Quantitative mass spectrometry in proteomics: critical review update from 2007 to the present. *Anal. Bioanal. Chem.* **404**(4), 939–965.

Barnidge, D. R., Hall, G. D., Stocker, J. L., and Muddiman, D. C. (2004). Evaluation of a cleavable stable isotope labeled synthetic peptide for absolute protein quantification using LC-MS/MS. *J. Proteome Res.* **3**(3), 658–661.

Bateman, R. H., Carruthers, R., Hoyes, J. B., Jones, C., Langridge, J. I., Millar, A., and Vissers, J. P. (2002). A novel precursor ion discovery method on a hybrid quadrupole orthogonal acceleration time-of-flight (Q-TOF) mass spectrometer for studying protein phosphorylation. *J. Am. Soc. Mass Spectrom.* **13**(7), 792–803.

Becher, D., Hempel, K., Sievers, S., Zühlke, D., Pané-Farré, J., Otto, A., Fuchs, S., Albrecht, D., Bernhardt, J., Engelmann, S., Völker, U., van Dijl, J. M., and Hecker, M. (2009). A proteomic view of an important human pathogen—towards the quantification of the entire *Staphylococcus aureus* proteome. *PLoS One* **4**(12), e8176.

Becher, D., Büttner, K., Moche, M., Hessling, B., and Hecker, M. (2011). From the genome sequence to the protein inventory of *Bacillus subtilis*. *Proteomics* **11**(15), 2971–2980.

Bensimon, A., Heck, A. J., and Aebersold, R. (2012). Mass spectrometry-based proteomics and network biology. *Annu. Rev. Biochem.* **81**, 379–405.

Bernhardt, J., Büttner, K., Scharf, C., and Hecker, M. (1999). Dual channel imaging of two-dimensional electropherograms in *Bacillus subtilis*. *Electrophoresis* **20**(11), 2225–2240.

Berth, M., Moser, F. M., Kolbe, M., and Bernhardt, J. (2007). The state of the art in the analysis of two-dimensional gel electrophoresis images. *Appl. Microbiol. Biotechnol.* **76**(6), 1223–1243.

Beynon, R. J., Doherty, M. K., Pratt, J. M., and Gaskell, S. J. (2005). Multiplexed absolute quantification in proteomics using artificial QCAT proteins of concatenated signature peptides. *Nat. Methods* **2**(8), 587–589.

Brownridge, P. and Beynon, R. J. (2011). The importance of the digest: proteolysis and absolute quantification in proteomics. *Methods* **54**(4), 351–360.

Brownridge, P., Holman, S. W., Gaskell, S. J., Grant, C. M., Harman, V. M., Hubbard, S. J., Lanthaler, K., Lawless, C., O'Cualain, R., Sims, P., Watkins, R., and Beynon, R. J. (2011). Global absolute quantification of a proteome: challenges in the deployment of a QconCAT strategy. *Proteomics* **11**(15), 2957–2970.

Brun, V., Dupuis, A., Adrait, A., Marcellin, M., Thomas, D., Court, M., Vandenesch, F., and Garin, J. (2007). Isotope-labeled protein standards: toward absolute quantitative proteomics. *Mol. Cell. Proteomics* **6**(12), 2139–2149.

Budde, I., Steil, L., Scharf, C., Völker, U., and Bremer, E. (2006). Adaptation of *Bacillus subtilis* to growth at low temperature: a combined transcriptomic and proteomic appraisal. *Microbiology* **152**(Pt. 3), 831–853.

Carrette, O., Burkhard, P. R., Sanchez, J. C., and Hochstrasser, D. F. (2006). State-of-the-art two-dimensional gel electrophoresis: a key tool of proteomics research. *Nat. Protoc.* **1**(2), 812–823.

Chuang, H. Y., Hofree, M., and Ideker, T. (2010). A decade of systems biology. *Annu. Rev. Cell Dev. Biol.* **26**, 721–744.

Cox, J. and Mann, M. (2007). Is proteomics the new genomics? *Cell* **130**(3), 395–398.

Cox, J. and Mann, M. (2011). Quantitative, high-resolution proteomics for data-driven systems biology. *Annu. Rev. Biochem.* **80**, 273–299.

de Godoy, L. M., Olsen, J. V., Cox, J., Nielsen, M. L., Hubner, N. C., Fröhlich, F., Walther, T. C., and Mann, M. (2008). Comprehensive mass-spectrometry-based proteome quantification of haploid versus diploid yeast. *Nature* **455**(7217), 1251–1254.

Desiere, F., Deutsch, E. W., Nesvizhskii, A. I., Mallick, P., King, N. L., Eng, J. K., Aderem, A., Boyle, R., Brunner, E., Donohoe, S., Fausto, N., Hafen, E., Hood, L., Katze, M. G., Kennedy, K. A., Kregenow, F., Lee, H., Lin, B., Martin, D., Ranish, J. A., Rawlings, D. J., Samelson, L. E., Shiio, Y., Watts, J. D., Wollscheid, B., Wright, M. E., Yan, W., Yang, L., Yi, E. C., Zhang, H., and Aebersold, R. (2005). Integration with the human genome of peptide sequences obtained by high-throughput mass spectrometry. *Genome Biol.* **6**(1), R9.

Domon, B. and Aebersold, R. (2010). Options and considerations when selecting a quantitative proteomics strategy. *Nat. Biotechnol.* **28**(7), 710–721.

Dreisbach, A., Otto, A., Becher, D., Hammer, E., Teumer, A., Gouw, J. W., Hecker, M., and Völker, U. (2008). Monitoring of changes in the membrane proteome during stationary phase adaptation of *Bacillus subtilis* using in vivo labeling techniques. *Proteomics* **8**(10), 2062–2076.

Dupuis, A., Hennekinne, J. A., Garin, J., and Brun, V. (2008). Protein standard absolute quantification (PSAQ) for improved investigation of staphylococcal food poisoning outbreaks. *Proteomics* **8**(22), 4633–4636.

Gallien, S., Duriez, E., and Domon, B. (2011). Selected reaction monitoring applied to proteomics. *J. Mass Spectrom.* **46**(3), 298–312.

Geiger, T., Cox, J., and Mann, M. (2010). Proteomics on an Orbitrap benchtop mass spectrometer using all-ion fragmentation. *Mol. Cell. Proteomics* **9**(10), 2252–2261.

Gerber, S. A., Rush, J., Stemman, O., Kirschner, M. W., and Gygi, S. P. (2003). Absolute quantification of proteins and phosphoproteins from cell lysates by tandem MS. *Proc. Natl. Acad. Sci. USA* **100**(12), 6940–6945.

Ghaemmaghami, S., Huh, W. K., Bower, K., Howson, R. W., Belle, A., Dephoure, N., O'Shea, E. K., and Weissman, J. S. (2003). Global analysis of protein expression in yeast. *Nature* **425**(6959), 737–741.

Gillet, L. C., Navarro, P., Tate, S., Rost, H., Selevsek, N., Reiter, L., Bonner, R., and Aebersold, R. (2012). Targeted data extraction of the MS/MS spectra generated by data-independent acquisition: a new concept for consistent and accurate proteome analysis. *Mol. Cell. Proteomics* **11**(6). doi:http://dx.doi.org/10.1074/mcp O111.016717.

Grossmann, J., Roschitzki, B., Panse, C., Fortes, C., Barkow-Oesterreicher, S., Rutishauser, D., and Schlapbach, R. (2010). Implementation and evaluation of relative and absolute quantification in shotgun proteomics with label-free methods. *J. Proteomics* **73**(9), 1740–1746.

Havlis, J. and Shevchenko, A. (2004). Absolute quantification of proteins in solutions and in polyacrylamide gels by mass spectrometry. *Anal. Chem.* **76**(11), 3029–3036.

Hecker, M., Antelmann, H., Büttner, K., and Bernhardt, J. (2008). Gel-based proteomics of Gram-positive bacteria: a powerful tool to address physiological questions. *Proteomics* **8**(23–24), 4958–4975.

Hecker, M., Reder, A., Fuchs, S., Pagels, M., and Engelmann, S. (2009). Physiological proteomics and stress/starvation responses in *Bacillus subtilis* and *Staphylococcus aureus*. *Res. Microbiol.* **160**(4), 245–258.

Huh, W. K., Falvo, J. V., Gerke, L. C., Carroll, A. S., Howson, R. W., Weissman, J. S., and O'Shea, E. K. (2003). Global analysis of protein localization in budding yeast. *Nature* **425**(6959), 686–691.

Ideker, T., Galitski, T., and Hood, L. (2001). A new approach to decoding life: systems biology. *Annu. Rev. Genomics Hum. Genet.* **2**, 343–372.

Jaquinod, M., Trauchessec, M., Huillet, C., Louwagie, M., Lebert, D., Picard, G., Adrait, A., Dupuis, A., Garin, J., Brun, V., and Bruley, C. (2012). Mass spectrometry-based absolute protein quantification: PSAQ strategy makes use of "noncanonical" proteotypic peptides. *Proteomics* **12**(8), 1217–1221.

Kettenbach, A. N., Rush, J., and Gerber, S. A. (2011). Absolute quantification of protein and post-translational modification abundance with stable isotope-labeled synthetic peptides. *Nat. Protoc.* **6**(2), 175–186.

Kirkpatrick, D. S., Gerber, S. A., and Gygi, S. P. (2005). The absolute quantification strategy: a general procedure for the quantification of proteins and post-translational modifications. *Methods* **35**(3), 265–273.

Kitano, H. (2002). Systems biology: a brief overview. *Science* **295**(5560), 1662–1664.

Lange, V., Picotti, P., Domon, B., and Aebersold, R. (2008). Selected reaction monitoring for quantitative proteomics: a tutorial. *Mol. Syst. Biol.* **4**, 222.

Luhn, S., Berth, M., Hecker, M., and Bernhardt, J. (2003). Using standard positions and image fusion to create proteome maps from collections of two-dimensional gel electrophoresis images. *Proteomics* **3**(7), 1117–1127.

Maass, S., Sievers, S., Zühlke, D., Kuzinski, J., Sappa, P. K., Muntel, J., Hessling, B., Bernhardt, J., Sietmann, R., Völker, U., Hecker, M., and Becher, D. (2011). Efficient, global-scale quantification of absolute protein amounts by integration of targeted mass spectrometry and two-dimensional gel-based proteomics. *Anal. Chem.* **83**(7), 2677–2684.

Maier, T., Schmidt, A., Guell, M., Kuhner, S., Gavin, A. C., Aebersold, R., and Serrano, L. (2011). Quantification of mRNA and protein and integration with protein turnover in a bacterium. *Mol. Syst. Biol.* **7**, 511.

Mallick, P., Schirle, M., Chen, S. S., Flory, M. R., Lee, H., Martin, D., Ranish, J., Raught, B., Schmitt, R., Werner, T., Küster, B., and Aebersold, R. (2007). Computational prediction of proteotypic peptides for quantitative proteomics. *Nat. Biotechnol.* **25**(1), 125–131.

Malmström, J., Beck, M., Schmidt, A., Lange, V., Deutsch, E. W., and Aebersold, R. (2009). Proteome-wide cellular protein concentrations of the human pathogen *Leptospira interrogans*. *Nature* **460**(7256), 762–765.

Martens, L., Hermjakob, H., Jones, P., Adamski, M., Taylor, C., States, D., Gevaert, K., Vandekerckhove, J., and Apweiler, R. (2005). PRIDE: the proteomics identifications database. *Proteomics* **5**(13), 3537–3545.

Mayya, V., Rezual, K., Wu, L., Fong, M. B., and Han, D. K. (2006). Absolute quantification of multisite phosphorylation by selective reaction monitoring mass spectrometry: determination of inhibitory phosphorylation status of cyclin-dependent kinases. *Mol. Cell. Proteomics* **5**(6), 1146–1157.

Michalik, S., Bernhardt, J., Otto, A., Moche, M., Becher, D., Meyer, H., Lalk, M., Schurmann, C., Schlueter, R., Kock, H., Gerth, U., and Hecker, M. (2012). Life and death of proteins: a case study of glucose-starved *Staphylococcus aureus*. *Mol. Cell. Proteomics* **11**(9), 558–570.

Michalski, A., Cox, J., and Mann, M. (2011). More than 100,000 detectable peptide species elute in single shotgun proteomics runs but the majority is inaccessible to data-dependent LC-MS/MS. *J. Proteome Res.* **10**(4), 1785–1793.

Middelberg, A. P. (1995). Process-scale disruption of microorganisms. *Biotechnol. Adv.* **13**(3), 491–551.

Mirzaei, H., McBee, J. K., Watts, J., and Aebersold, R. (2008). Comparative evaluation of current peptide production platforms used in absolute quantification in proteomics. *Mol. Cell. Proteomics* **7**(4), 813–823.

Neidhardt, F. C. (2011). How microbial proteomics got started. *Proteomics* **11**(15), 2943–2946.

Olsen, J. V., Ong, S. E., and Mann, M. (2004). Trypsin cleaves exclusively C-terminal to arginine and lysine residues. *Mol. Cell. Proteomics* **3**(6), 608–614.

Otto, A., Bernhardt, J., Meyer, H., Schaffer, M., Herbst, F. A., Siebourg, J., Mäder, U., Lalk, M., Hecker, M., and Becher, D. (2010). Systems-wide temporal proteomic profiling in glucose-starved *Bacillus subtilis*. *Nat. Commun.* **1**, 137.

Panchaud, A., Scherl, A., Shaffer, S. A., von Haller, P. D., Kulasekara, H. D., Miller, S. I., and Goodlett, D. R. (2009). Precursor acquisition independent from ion count: how to dive deeper into the proteomics ocean. *Anal. Chem.* **81**(15), 6481–6488.

Panchaud, A., Jung, S., Shaffer, S. A., Aitchison, J. D., and Goodlett, D. R. (2011). Faster, quantitative, and accurate precursor acquisition independent from ion count. *Anal. Chem.* **83**(6), 2250–2257.

Picotti, P. and Aebersold, R. (2012). Selected reaction monitoring-based proteomics: workflows, potential, pitfalls and future directions. *Nat. Methods* **9**(6), 555–566.

Picotti, P., Rinner, O., Stallmach, R., Dautel, F., Farrah, T., Domon, B., Wenschuh, H., and Aebersold, R. (2009). High-throughput generation of selected reaction-monitoring assays for proteins and proteomes. *Nat. Methods* **7**(1), 43–46.

Poetsch, A., Haussmann, U., and Burkovski, A. (2011). Proteomics of corynebacteria: from biotechnology workhorses to pathogens. *Proteomics* **11**(15), 3244–3255.

Pratt, J. M., Simpson, D. M., Doherty, M. K., Rivers, J., Gaskell, S. J., and Beynon, R. J. (2006). Multiplexed absolute quantification for proteomics using concatenated signature peptides encoded by QconCAT genes. *Nat. Protoc.* **1**(2), 1029–1043.

Rabilloud, T., Vaezzadeh, A. R., Potier, N., Lelong, C., Leize-Wagner, E., and Chevallet, M. (2009). Power and limitations of electrophoretic separations in proteomics strategies. *Mass Spectrom. Rev.* **28**(5), 816–843.

Rabilloud, T., Chevallet, M., Luche, S., and Lelong, C. (2010). Two-dimensional gel electrophoresis in proteomics: past, present and future. *J. Proteomics* **73**(11), 2064–2077.

Rivers, J., Simpson, D. M., Robertson, D. H. L., Gaskell, S. J., and Beynon, R. J. (2007). Absolute multiplexed quantitative analysis of protein expression during muscle development using QconCAT. *Mol. Cell. Proteomics* **6**(8), 1416.

Schmidt, A., Claassen, M., and Aebersold, R. (2009). Directed mass spectrometry: towards hypothesis-driven proteomics. *Curr. Opin. Chem. Biol.* **13**(5–6), 510–517.

Schmidt, A., Beck, M., Malmström, J., Lam, H., Claassen, M., Campbell, D., and Aebersold, R. (2011). Absolute quantification of microbial proteomes at different states by directed mass spectrometry. *Mol. Syst. Biol.* **7**, 510.

Schwanhäusser, B., Busse, D., Li, N., Dittmar, G., Schuchhardt, J., Wolf, J., Chen, W., and Selbach, M. (2011). Global quantification of mammalian gene expression control. *Nature* **473**(7347), 337–342.

Silva, J. C., Denny, R., Dorschel, C. A., Gorenstein, M., Kass, I. J., Li, G. Z., McKenna, T., Nold, M. J., Richardson, K., Young, P., and Geromanos, S. (2005). Quantitative proteomic analysis by accurate mass retention time pairs. *Anal. Chem.* **77**(7), 2187–2200.

Silva, J. C., Denny, R., Dorschel, C., Gorenstein, M. V., Li, G. Z., Richardson, K., Wall, D., and Geromanos, S. J. (2006a). Simultaneous qualitative and quantitative analysis of the *Escherichia coli* proteome: a sweet tale. *Mol. Cell. Proteomics* **5**(4), 589–607.

Silva, J. C., Gorenstein, M. V., Li, G. Z., Vissers, J. P., and Geromanos, S. J. (2006b). Absolute quantification of proteins by LCMSE: a virtue of parallel MS acquisition. *Mol. Cell. Proteomics* **5**(1), 144–156.

Soufi, B., Kumar, C., Gnad, F., Mann, M., Mijakovic, I., and Macek, B. (2010). Stable isotope labeling by amino acids in cell culture (SILAC) applied to quantitative proteomics of *Bacillus subtilis*. *J. Proteome Res.* **9**(7), 3638–3646.

Tal, M., Silberstein, A., and Nusser, E. (1985). Why does Coomassie Brilliant Blue R interact differently with different proteins? A partial answer. *J. Biol. Chem.* **260**(18), 9976–9980.

Ünlü, M., Morgan, M. E., and Minden, J. S. (1997). Difference gel electrophoresis: a single gel method for detecting changes in protein extracts. *Electrophoresis* **18**(11), 2071–2077.

Venable, J. D., Dong, M. Q., Wohlschlegel, J., Dillin, A., and Yates, J. R. (2004). Automated approach for quantitative analysis of complex peptide mixtures from tandem mass spectra. *Nat. Methods* **1**(1), 39–45.

Völker, U. and Hecker, M. (2005). From genomics via proteomics to cellular physiology of the Gram-positive model organism *Bacillus subtilis*. *Cell. Microbiol.* **7**(8), 1077–1085.

Wenzel, M. and Bandow, J. E. (2011). Proteomic signatures in antibiotic research. *Proteomics* **11**(15), 3256–3268.

Westermeier, R. and Marouga, R. (2005). Protein detection methods in proteomics research. *Biosci. Rep.* **25**(1–2), 19–32.

Wolff, S., Otto, A., Albrecht, D., Zeng, J. S., Büttner, K., Glückmann, M., Hecker, M., and Becher, D. (2006). Gel-free and gel-based proteomics in *Bacillus subtilis*: a comparative study. *Mol. Cell. Proteomics* **5**(7), 1183–1192.

Zorn, M. E., Gibbons, R. D., and Sonzogni, W. C. (1997). Weighted least-squares approach to calculating limits of detection and quantification by modeling variability as a function of concentration. *Anal. Chem.* **69**(15), 3069–3075.

CHAPTER 4

Imaging fluorescent protein fusions in live bacteria

Geoff Doherty[1], Karla Mettrick[1], Ian Grainge, Peter J. Lewis[2]

School of Environmental and Life Sciences, University of Newcastle, Callaghan, New South Wales, Australia
[2]*Corresponding author. e-mail address: Peter.Lewis@newcastle.edu.au*

1 INTRODUCTION

Although they lack the defined subcellular organelles that eukaryotes use to partition cellular processes, it has become evident over the past 15 years that distinct subcellular domains are present in prokaryotes. The majority of our knowledge of subcellular compartmentalisation has arisen through the use of fluorescent tags such as GFP and its derivatives. The approaches utilised to identify protein localisation and abundance are suitable for high-throughput systems biology projects (Meile et al., 2006; Werner et al., 2009; Buescher et al., 2012). Here, we describe some of the molecular toolkits currently available to observe the subcellular localisation of both fully and partially functional fusions in a variety of prokaryotes, focussing on *Bacillus subtilis*, *Escherichia coli*, *Staphylococcus aureus* and *Acinetobacter baylyi*.

2 MOLECULAR TOOLKITS

There are a variety of tools available for labelling proteins in live, intact cells. Most of the tools comprise gene fusions constructed on a plasmid that can be either integrative or non-integrative, which are then transformed into the target cell. Alternatively, simple PCR constructs can be used to create gene fusions relying upon homologous recombination for integration. With an explosion of fluorescent tags having entered the cytological ball park since Martin Chalfie first described the use of GFP from jellyfish *Aequorea victoria* (Chalfie et al., 1994), researchers have a myriad of choices when undertaking localisation studies. These include GFP variants that have improved biophysical properties and shifted spectra (Chalfie et al., 1994; Heim et al., 1995; Cormack et al., 1996; Heim and Tsien, 1996), and also include the more recently developed mFruits, which are monomeric variants of the *Discosoma* genus of coral fluorescent protein dsRed (Shu et al., 2006).

---

[1]Contributed equally to this work.

These advances have allowed researchers to visualise several proteins simultaneously within the same live cell. In this section, we describe some of the considerations needed when undertaking fluorescent protein labelling experiments and highlight some of the more successfully utilised molecular systems used to label proteins in B. subtilis, E. coli, S. aureus and A. baylyi.

2.1 Bacillus subtilis

B. subtilis is one of the more highly utilised organisms used to study fluorescent protein localisation. There are numerous integrative plasmid systems available that can give rise to either N- or C-terminal fusions, and they can be designed to integrate into the chromosome at the target gene locus via a single-crossover integration, or to integrate at a different locus via a double-crossover integration. Examples of some popular B. subtilis vectors are shown in Figure 4.1 and Table 4.1. Single-crossover integration involves the entire plasmid integrating into the chromosome and results in the expression of C-terminal fluorescent protein fusions being driven by the target gene's natural promoter, as shown schematically in Figure 4.2. This gives rise to a fusion product with wild-type expression levels, which not only allows highly accurate subcellular localisation data to be determined but also facilitates rapid quantification of protein abundance (Doherty et al., 2010; Buescher et al., 2012). These plasmids typically contain auxiliary promoters such as IPTG or xylose-inducible promoters to drive expression of genes downstream of the target gene in the event they are located within an operon.

The *amyE* locus is a common choice for double-crossover integration in B. subtilis. Integration vectors targeting this locus contain both 5'- and 3'-regions of homology to *amyE* and are designed to express the gene fusion from either an IPTG or a xylose-inducible promoter. A schematic illustrating a double-crossover integration into the chromosome is shown in Figure 4.2. Unlike single-crossover vectors that are well suited for the production of C-terminal fusions, double-crossover vectors can also be used for the generation of either N- or C-terminal fluorescent protein fusions (Figure 4.1; Lewis and Marston, 1999; Feucht and Lewis, 2001). Integration via double crossover at *amyE* can be screened by inactivation of the *amyE* gene as outlined below. This type of system is useful in the event the protein fusion is partially functional. For example, the partially functional fusions to the essential cell division protein FtsZ expressed from the *amyE* locus can still be used to give accurate localisation data while it is functionally complemented by the wild-type copy of the gene (Feucht and Lewis, 2001).

Numerous examples of co-localisation of two proteins exist in B. subtilis using CFP/YFP or GFP/mCherry combinations (Lewis et al., 2000; Doherty et al., 2010). Because integrative plasmids rely on regions of homology to integrate efficiently into the chromosome, homology between their plasmid backbones can complicate strain construction. Consequently, care needs to be taken when designing such dual-labelling experiments. An effective plasmid combination for dual-labelling experiments that avoids this problem is that of the *gfpmut3* containing pYG1 and the *mCherry* containing pNG621 (Figure 4.1 and Table 4.1; Doherty et al., 2010). These two plasmids share little sequence homology, which reduces the chance of

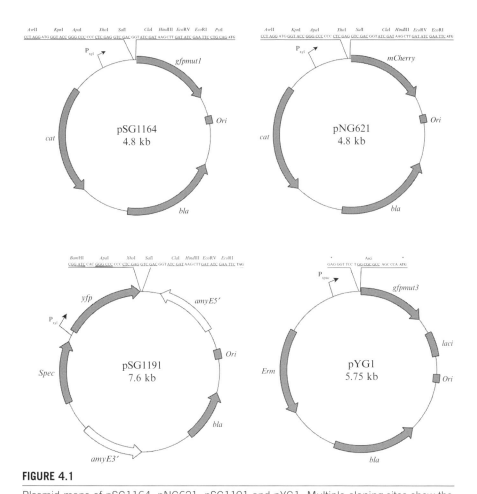

FIGURE 4.1

Plasmid maps of pSG1164, pNG621, pSG1191 and pYG1. Multiple cloning sites show the reading frame and all unique restriction enzyme sites are underlines. Stop and start codons are bold. Asterisks in pYG1 show the nucleotides that the digestion stops at during T4 DNA polymerase reaction when dTTP is added to the reaction. *bla*, *cat*, *Erm* and *Spec*, ampicillin, chloramphenicol, erythromycin and spectinomycin resistance, respectively; *Ori*, plasmid origin of replication; P_{xyl}, xylose-inducible promoter; P_{spac}, IPTG-inducible promoter. Plasmids not to scale.

recombination occurring between plasmid-derived sequences. Furthermore, they contain compatible antibiotic resistance genes enabling efficient strain construction.

2.1.1 Cloning into pSG- and pNG-based vectors (Table 4.1)

1. PCR amplify the gene of interest using primers that include restriction enzyme sites chosen from the plasmid's multiple cloning site (MCS). Confirm that these sites are not present *within* the amplified PCR products. Purify the resulting PCR product using any available PCR cleanup kit (refer to Figure 4.1 for restriction sites).

Table 4.1 Plasmids used for the construction of fluorescent protein fusions in a variety of microbial species

Plasmid	Description	References
pKD46[a]	λ Red recombinase expression	Datsenko and Wanner (2000)
pCP20[a]	FLP-recombinase expression	Datsenko and Wanner (2000)
pKD3[a]	Contains chloramphenicol resistance cassette flanked by FLP sequences	Datsenko and Wanner (2000)
pKD4[a]	Contains kanamycin resistance cassette flanked by FLP sequences	Datsenko and Wanner (2000)
pSG1164[b,c]	C-terminal GFPmut1 fusion for single-crossover integration with P_{xyl} driving downstream genes	Lewis and Marston (1999)
pSG1154[b]	C-terminal GFPmut1 fusion for double-crossover integration at amyE driven by P_{xyl}	Lewis and Marston (1999)
pSG1186[b,c]	C-terminal CFP fusion for single-crossover integration	Feucht and Lewis (2001)
pSG1187[b,c]	C-terminal YFP fusion for single-crossover integration	Feucht and Lewis (2001)
pSG1729[b]	N-terminal GFPmut1 fusion for double-crossover integration at amyE driven by P_{xyl}	Lewis and Marston (1999)
pSG1190[b]	N-terminal CFP fusion for double-crossover integration at amyE driven by P_{xyl}	Feucht and Lewis (2001)
pSG1191[b]	N-terminal YFP fusion for double-crossover integration at amyE driven by P_{xyl}	Feucht and Lewis (2001)
pSG1192[b]	C-terminal CFP fusion for double-crossover integration at amyE driven by P_{xyl}	Feucht and Lewis (2001)
pSG1193[b]	C-terminal YFP fusion for double-crossover integration at amyE driven by P_{xyl}	Feucht and Lewis (2001)
pYG1[b]	C-terminal GFPmut3 fusion for single-crossover integration with P_{xyl} driving downstream genes	Doherty et al. (2010)
pNG621[b,c]	C-terminal mCherry fusion for single-crossover integration with P_{xyl} driving downstream genes	Doherty et al. (2010)
pLOW-GFP[d]	C-terminal GFPmut1 fusion. Low copy, non-integrative	Liew et al. (2011)
pLau53[a]	C-terminal EYFP and ECFP fusion to tetR and lacI, respectively. Tandem expression driven by P_{ara}	Lau et al. (2003)
Primers	λ Red Template construction	
GFP F	TGGCGCGCTCTAGAAATGCGTAAAG	Doherty et al. (2010)
GFP R[e]	CCCCTCGGTACCTTATTTGTATAG	Doherty et al. (2010)
Cat F[e]	AAAAAAGGTACCGTGTAGGCTGGAGCTGCTTC	Doherty et al. (2010)
Cat R	AAAAAATCATGAATGGGAATTAGCCATGGTCC	Doherty et al. (2010)

[a]For use in E. coli.
[b]For use in Bacillus spp.
[c]For use in Acinetobacter spp.
[d]For use in S. aureus.
[e]Acc651 restriction sites used for joining gfpmut3 to chloramphenicol resistance gene are underlined.

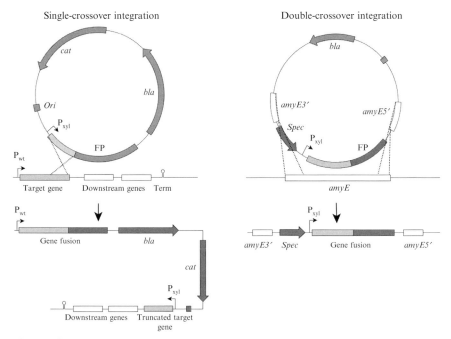

FIGURE 4.2

Schematics of single- and double-crossover integration. Homologous recombination between the truncated gene sequence on the plasmid and target gene on the chromosome results the fusion of the gene encoding the fluorescent protein to the full-length target gene driven by the wild-type promoter. In this example, the xylose-inducible promoter (P_{xyl}) allows expression of the truncated portion of the gene as well as the expression of any genes that lie downstream in the operon. During double-crossover integration, homologous recombination between the 5'- and 3'-portions of *amyE* on the plasmid with the full-length *amyE* gene in the chromosome results in the insertion of the gene fusion, along with a resistance marker (Spec). This results in the fusion being driven by the xylose-inducible promoter and the inactivation of *amyE*. Plasmid labels as per Figure 4.1.

   a. For C-terminal fluorescent protein fusions, ensure the stop codon is engineered out of the reverse primer. For N-terminal fusions, ensure the stop codon is retained.
   b. For fully functional gene fusions, around 500 bp of the 3'-end of the gene should be amplified. In the case of genes smaller than this, clone the entire gene without the start codon.
2. Digest the PCR product and plasmid with appropriate restriction enzymes for 2 h at 37 °C before inactivating either by 65 °C for 30 min or by use of a PCR cleanup kit.
3. Mix 1 μL of ligase (e.g. 5 units/μL; Fermentas), 2 μL of 10 × ligase buffer and approximately 1 μg of digested vector and PCR insert, and make up to 20 μL with water. The relative ratio of vector to insert should be ∼1:3. Incubate at room temperature for 2 h or overnight at 4 °C.

4. Mix 50 μL of competent *E. coli* DH5α cells (prepared by the method of Hanahan, 1983) with 5 μL ligation mixture and incubate on ice for 1 h. Pulse heat at 42 °C for 90 s before adding 150 μL of LB. Incubate at 37 °C for 1 h and then plate onto nutrient agar containing the appropriate antibiotic.
5. Screen colonies using a forward primer that anneals 50 nt upstream and a reverse primer that anneals 50 nt downstream from the MCS or ligation-independent cloning (LIC) site. This will generate a band of 100 bp if cloning is unsuccessful and a band of the gene fragment size plus 100 bp if cloning is successful. The positions of these primers also make them ideal for sequencing the inserted DNA once positive clones have been identified.

2.1.2 Cloning into pYG1

1. Cloning into pYG1 using LIC requires generation of long single-stranded overhangs, and consequently, PCR primers containing 5′-GGGTTCCTGGCGCGAGC-3′ at the 5′-end of the forward primer and 5′-TTGGGCTGGCGCGAGC-3′ at the 5′-end of the reverse primer must be used. Purify the resulting PCR product using any available PCR cleanup kit (refer to Doherty *et al.*, 2010, e.g. primers).
2. Digest pYG1 with *Asc*I for 2 h at 37 °C to linearise.
3. Incubate 0.3 pmol of insert with T4 DNA polymerase (1 unit; NEB) and 2.5 mM dTTP for 30 min at room temperature. Incubate 0.3 pmol of linearised vector with T4 DNA polymerase (1 unit; NEB) and 2.5 mM dATP for 30 min at room temperature.
4. Mix vector and insert in a ratio of ~1:3 for 10 min at room temperature, transform into *E. coli* and screen as above.

2.1.3 Transforming B. subtilis 168

Solutions

10 × PC	
21.4 g	K_2HPO_4
12 g	KH_2PO_4
2 g	$Na_3Citrate \cdot 5H_2O$
1.7 g	$Na_3Citrate \cdot 2H_2O$

Make up to 1 L with dH_2O and autoclave

MD medium	
5 mL	10 × PC
2.5 mL	40% glucose (w/v)
1.25 mL	trp (2 mg/mL)
0.25 mL	Ferric ammonium citrate (2.2 mg/mL)
2.5 mL	L-Aspartate (50 mg/mL)
0.15 mL	1 M $MgSO_4$

Make up to 50 mL in dH$_2$O, pH to 7.0 with KOH and filter sterilise

1. Inoculate 10 mL of MD medium supplemented with 50 µL of 20% (w/v) of casamino acids with *B. subtilis* to an OD$_{600}$ of 0.2 from a freshly struck plate and incubate at 37 °C with shaking until OD$_{600}$ of 1–1.5 is reached.
2. Add an equal volume of MD medium and continue shaking for a further 1 h at 37 °C.
3. Add 1 ng of purified plasmid DNA to 800 µL of the culture and shake for 20 min at 37 °C.
4. Add 25 µL of 20% (w/v) casamino acids and shake for a further 1–1.5 h at 37 °C and then plate onto nutrient agar supplemented with appropriate antibiotic.

2.1.4 Screening transformants

Screening can typically be done by checking the fluorescence of the transformed cells and also by the following:

a. Check the C-terminal single-crossover constructs by PCR, using a forward primer that anneals upstream of the forward primer used in the initial PCR amplification of the gene and the reverse primer used in the screening process (anneals to the plasmid backbone). An example would be to use a forward primer that anneals to the promoter region of the target gene and a reverse primer that anneals to the gene encoding the fluorescent protein.
b. In the case of double-crossover constructs into the *amyE* locus, screen for the hydrolysis of starch by patching colonies onto a nutrient agar plate containing 1% starch and incubating overnight. Patch a positive control in a defined location (amylase positive). In the morning, place a few iodine crystals on the inverted lid and allow them to vaporise onto the surface of the agar for ~30 s. Double-crossover clones will have inactivated *amyE*, do not hydrolyse starch and therefore do not have a halo around the colonies.

2.2 *Escherichia coli*

Another organism that has been widely studied using fluorescent proteins is *E. coli*. As with *B. subtilis*, the ideal situation is to create a functional fluorescent fusion protein expressed from the natural endogenous promoter. However, a plethora of non-integrative plasmids exists such as pBAD and pBluescript that can be modified to contain a fluorescent protein gene that is suitable for the generation of fusions driven by any number of inducible promoters. These systems are analogous to the double-crossover integrative plasmids mentioned above for *B. subtilis*, in that gene expression is controlled from an inducible promoter, and the intact wild-type copy is available for complementation of any partially functional fusions. The majority of common plasmid cloning vectors are maintained at a high copy number in *E. coli*. This can make it more difficult to control expression levels precisely. Therefore, the choice of promoter used to drive expression of the fusion gene, as well as plasmid copy number, should be taken into account when considering an appropriate level of fluorescent protein expression.

One of the more elegant approaches in strain construction is the adaptation of the λ Red system originally developed to create markerless gene knockout strains in *E. coli* (Datsenko and Wanner, 2000). We have successfully used this system to create markerless GFP and mCherry strains in *E. coli*, which contain gene fusions driven by their wild-type promoters (Doherty et al., 2010). Briefly, a PCR product containing the fluorescent tag and antibiotic resistance cassette flanked by homology to the gene of interest is transformed into electrocompetent *E. coli* expressing the λ Red genes, and recombinants are selected for using the antibiotic resistance. This resistance cassette is flanked by FLP-recombinase sequences (FRT), which allows subsequent removal when transformed with pCP20, which encodes the FLP recombinase (Table 4.1). The protocol for creating fluorescently labelled *E. coli* strains using the λ Red system is outlined below and schematically in Figure 4.3. We illustrate the process using the chloramphenicol resistance gene from pKD3, but the system works equally well using the kanamycin resistance gene from pKD4. (NB. pKD46 and pCP20 are temperature sensitive plasmids and will be lost from the cells when grown at 37 °C without ampicillin selection.)

1. PCR amplify the chloramphenicol resistance gene (Figure 4.3 light grey box) flanked by the FLP-recombinase sequences (Figure 4.3 grey striped boxes) using pKD3 as the template. Separately, amplify the desired fluorescent protein (Figure 4.3 dark grey box) from any available template. The forward primer used for amplifying the antibiotic resistance gene and the reverse primer for amplifying the fluorescent protein-encoding gene are engineered with the same restriction site (or restriction sites with compatible ends) to allow ligation. Refer to Table 4.1 for primers used to create the *gfpmut3*-chloramphenicol resistance cassette.
2. Clean up both the PCR products using a standard PCR cleanup kit, digest with appropriate restriction enzyme(s) and ligate. Using the fluorescent protein forward primer and the chloramphenicol acetyl transferase reverse primer, perform PCR on the above ligation and gel purify the ∼1800-bp product. This can be used as the template for creating any C-terminal protein fusion.
3. Design primers that will amplify the above template that also contain the last 50 nt of the gene of interest (not including the stop codon) and 50 nt downstream of the stop codon. This provides enough homology for the recombinase to integrate into the chromosome.
4. Transform the desired *E. coli* strain with pKD46, the λ Red recombinase expressing plasmid, and plate onto nutrient agar containing 100 μg/mL ampicillin and grow at 30 °C overnight. From a single colony, grow at 30 °C to an OD_{600} of 0.1 in LB containing 100 μg/mL ampicillin before adding L-arabinose to a final concentration of 10 mM to induce λ Red recombinase expression. Grow until an OD_{600} of 0.4 is reached and then make the cells electrocompetent using standard procedures. Now, electroporate the PCR construct into the pKD46-induced electrocompetent cells and plate onto nutrient agar plates containing chloramphenicol (20 μg/mL) and incubate overnight at 37 °C (see below for an electroporation protocol). Once recombination in the strain has been confirmed

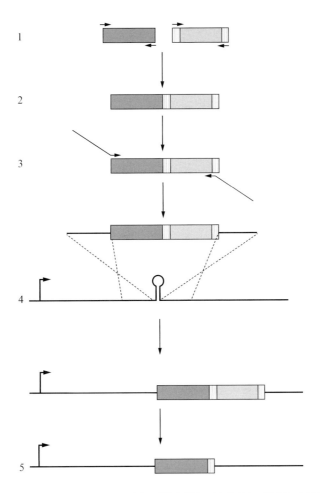

FIGURE 4.3

Construction of the markerless fluorescent protein fusion strains using the λ Red system. 1. PCR amplify the chloramphenicol or kanamycin resistance gene (light grey box) flanked by the FLP-recombinase sequences (grey striped boxes) and the desired fluorescent protein (dark grey box). 2. Clean up, digest and ligate both PCR products. Using the flanking primers, PCR amplify and gel purify the entire ∼1800 bp product. 3. Create a fluorescent protein fusion by PCR amplification from the above template using primers that contain at least 50 nt of homology to the desired insertion site on the chromosome. 4. Transform into pKD46-induced electrocompetent cells where recombination occurs, directed by the two 50 bp homology regions 5. Make this strain electrocompetent and transform with pCP20 to remove antibiotic resistance using FLP-recombinase.

by PCR, patch onto nutrient agar plates with and without ampicillin to confirm loss of pKD46.
5. Make the new strain electrocompetent, transform with pCP20 and plate onto nutrient agar containing 100 µg/mL ampicillin and grow at 30 °C overnight. From a single colony, grow at 30 °C to an OD_{600} of 0.5 in LB before inducing FLP-recombinase expression by shifting the temperature to 43 °C for 1 hour, then plate onto nutrient agar plates (serial dilutions will be needed to ensure single colonies are obtained). To confirm loss of antibiotic resistance and the loss of pCP20, patch onto nutrient agar without any antibiotics, with ampicillin and with chloramphenicol. Select a colony with no antibiotic resistance and check microscopically for fluorescence.

2.3 Staphylococcus aureus

S. aureus is one of the more recalcitrant bacterial species with respect to genetic modification due to the presence of multiple restriction and modification systems. A recently developed non-integrative low-copy plasmid system has been utilised for protein localisation studies and represents a major breakthrough in studying subcellular protein dynamics in this important pathogen (Liew et al., 2011). This system uses a low-copy number plasmid (pLow) with a tightly controlled IPTG-inducible promoter to provide precise control of the level of protein fusion produced, and is currently the most efficient way to fluorescently label proteins in this organism. pLow is a shuttle plasmid that can be grown in *E. coli* for cloning and amplification purposes prior to electroporation into *S. aureus* using the approach described by Grkovic et al. (2003).

2.4 Acinetobacter spp.

Acinetobacter spp. are fast emerging as very important clinical pathogens and are the focus of increased research. *A. baylyi* ADP1 could be classified as a super-competent organism and can be transformed with recombinant DNA by simply adding it to an exponentially growing culture (de Berardinis et al., 2008). *A. baumannii* transformation can be achieved by standard electroporation procedures such as that outlined below.

1. From a single colony, grow overnight at 37 °C in 2 mL of LB
2. In the morning, make a 1:50 dilution in 10 mL of LB and grow to an OD_{600} of 0.5
3. Remove cells to a fresh 50-mL Falcon-type tube and chill on ice for 10 min before pelleting cells
4. Wash the pellet three times in 5 mL of chilled dH_2O, before resuspending in 400 µL of dH_2O
5. Mix 1 ng of DNA to be transformed with 50 µL of cells in a 0.2-cm electroporation cuvette and electroporate according to manufacturer's instructions (settings for BioRad MicroPulser are 25 mFD, 200 W and 2.5 kV)

We have created C-terminal fluorescent gene fusions by single crossover with integrative plasmids such as those designed for *B. subtilis* (Table 4.1). From our experience, at least 500 bp of homology is required to ensure efficient recombination.

2.5 Tet array and pLau53

Fluorescent repressor-operator systems were first developed to visualise specific DNA sites in eukaryotes (Robinett et al., 1996) but have been successfully used to study chromosome segregation in E. coli, B. subtilis and Caulobacter crescentus (Gordon et al., 1997; Webb et al., 1998; Viollier et al., 2004). The system has since been developed further to include dual expression of fluorescent proteins (Lau et al., 2003) as well as creating road-blocks for the study of DNA-replication fork stalling (Possoz et al., 2006). These systems contain tandem copies of *lacO* and/or *tetO* arrays inserted into specific chromosome regions. Originally, 256 copies were inserted into the chromosome although currently as few as 64 have been used successfully (Fekete and Chattoraj, 2005). A plasmid (pLau53) producing C-terminal CFP and YFP fusions of LacI and TetR, respectively, under the control of the arabinose promoter from pBAD has been constructed. The repressors bind specifically to their cognate operators and facilitate visualisation of the loci (at least two) by fluorescence microscopy. This system also allows for time-lapse analysis to visualise chromosome dynamics in living cells.

3 FUNCTIONAL ANALYSIS

Once a gene fusion has been created, it is critically important to assess the functionality of the product. In cases in which the target gene is essential for growth, this can be as simple as determining the growth kinetics with respect to wild-type strains (e.g. Davies et al., 2005). For non-essential genes, it is important to investigate the conditions that would allow an assessment of the functionality. For example, a gene essential for arabinose assimilation should be assessed for growth on arabinose as the sole carbon source. Genes involved in stress responses should likewise be assessed under the conditions that would allow the functionality to be determined. In the case of transcription factors, there are numerous *lacZ* fusions available to ensure that wild-type gene regulation is occurring in the recombinant strain. Finally, cell morphology should be assessed under the microscope to ensure that the size and shape of the cells are indistinguishable from the wild-type strain. A recent article illustrating aggregation artefacts that can arise from fluorescent protein fusions illustrates the importance of thoroughly investigating the functionality of the fusion product to ensure costly and time-consuming experiments are not performed on non-functional fusions (Landgraf et al., 2012).

4 GROWTH CONDITIONS

There are a variety of both liquid and solid growth media available to researchers. Generally speaking, the richer the medium, the larger the cells and therefore more spatial segregation within the cell. This is certainly the case when studying transcription and translation machineries, which occupy very clear subcellular domains during rapid growth in rich media, compared to that seen in minimal media (Lewis et al., 2000). Consistency is the key to obtaining clear reproducible images. This is of particular importance when comparing fluorescently labelled proteins from different strains. The media, the glassware and the incubation parameters all need to be standardised in order to be able to compare with confidence how two labelled proteins behave in different strains (Buescher et al., 2012).

4.1 Liquid

Under most circumstances, cells being prepared for microscopy can be grown in LB at 37 °C with aeration. Microscopy of stationary phase cells can be done using cells grown in overnight cultures with any required supplements. For microscopy of exponential growing cells, overnight cultures should be diluted back to an OD_{600} of 0.01 and grown with aeration to an optical density of approximately 0.3. One way to reduce waiting time and maximise the time spent on the microscope is to culture cells in 96-well microplates. This allows up to 12 different strains to be comfortably imaged by using simple dilutions. Fill the wells along Row A of the microplate with 200 μL of growth medium and then add 100 μL to all other wells. Add 2 μL of overnight culture to the 200 μL in Row A, mix thoroughly and take 100 μL to mix with the contents of Row B. Continue this 1:2 dilution down the plate to Row H and discard the final 100 μL. This gives eight dilutions of up to 12 cultures, providing flexibility in imaging without the use of excessive glassware and incubator space.

Cells carrying pLau53 require induction at an OD_{600} of ∼0.05 with 0.05% (w/v) L-arabinose and grown for a further 1–3 h. Cells that require lower levels of fluorescent protein production can be induced with 0.01% (w/v) L-arabinose at an OD_{600} of 0.2–0.6. Variability will occur with respect to the requirements for induction and should be optimised for each strain and circumstance.

If visualisation of the nucleoid is required, 5 μL of DAPI (4′,6-diamidino-2-phenylindole; 30 ng/mL stock) is added to 5 μL of the culture immediately prior to being placed on the slide.

4.2 Solid

Cells can be grown on a slide mounted with an agarose-based medium (1.2% agarose, w/v, in minimal media, e.g. M9). We have found that several fusions lose their localisation patterns when cells are attached to poly-L-lysine-coated slides, and consequently, we routinely use agarose-based slides for all imaging of fluorescent protein fusions. Rich media such as LB are not recommended due to high levels of autofluorescence although this artefact can be removed to some extent as described below under image processing. A temperature-controlled stage may be required. Other considerations include humidity, pH, oxygen, photobleaching and phototoxicity. Time-lapse microscopy can then be used by taking successive frames at specified time intervals (e.g. 30 min).

5 AGAROSE SLIDE PREPARATION

An agarose pad on a microscope slide creates a flat surface that is essential for high-quality imaging and ensures the entire field is in focus. The agarose also draws in the liquid from the culture resulting in the cells being embedded on the agarose surface, free of any movement yet able to grow and spread two-dimensionally on the surface.

- Pipette 500 µL of molten 1.2% agarose (w/v in H_2O or medium) onto a level glass microscope slide.
- Immediately, drop a coverslip carefully onto the agarose. Hold the coverslip at a 45° angle, allow the bottom edge to just touch the molten agarose and then gently let the coverslip drop on to the surface so that no air bubbles are trapped. An alternative is to use a Gene Frame®. These are adhesive plastic frames that can be stuck to a slide to create a space into which a small volume of molten agarose can be pipetted (the volume depends on the size of the frame used). The molten agarose can then be immediately overlayed with the Gene Frame® cover. When set, the coverslip can be carefully peeled off leaving a large flat surface. Due to their low volume, agarose pads created this way dry out very quickly unless kept in a humidified chamber.
- Once solidified, either proceed to applying a sample of bacterial culture or, for use later in the day, slides may be kept in the fridge wrapped in damp tissue.
- Gently remove the coverslip and dispense ~ 5 µL of liquid grown culture onto the centre of the pad. (NB. If cells are at a low OD, a larger volume can be spun down and resuspended to generate a cell density that is easy to visualise. Cells can also be counterstained to visualise other components, for example, DAPI for nucleoids, FM-64 (or similar) for cell membranes (Johnson *et al*., 2004)).
- Once the liquid has dried, place a fresh coverslip onto the pad.

6 IMAGING HARDWARE

6.1 Microscope and camera

For epifluorescence imaging, a suitable microscope (either upright or inverted) equipped with a Peltier-cooled charge-coupled device (CCD) camera is required. There are many highly sensitive digital camera systems designed for this purpose, although we currently use the Hamamatsu Orca-ER. Due to the small size of bacteria and the often relatively low cellular levels of the fluorescent protein fusion under investigation, the requirements of the microbial cell biologist are different to those used to imaging eukaryotic cells. Signal intensity is low and cells are prone to photo-damage if long exposure times are used. Fluorescence is rapidly quenched, meaning that minimal illumination should be used prior to image acquisition.

In order to detect the low emission levels from fluorescent proteins, the light source must be powerful. Often used are high-energy short arc-discharge lamps such as mercury burners (50–200 W) or xenon burners (75–150 W). These lamps should not be used beyond their recommended lifetime due to loss of efficiency and an increased risk of shattering. Whilst the imaging system may claim to have a broad spectral range (e.g. 300–900 nm), the quantum efficiency will only be high over a limited range. Many of the 'bioimaging' CCDs have been modified so that the quantum efficiency is enhanced at the shorter wavelengths (blue/green). CCDs are traditionally most sensitive in the far-red/IR range.

Current high-resolution cameras will have a pixel size in the region of 6.45×6.45 µm. While high resolution is important, there is always a balance that must be achieved between resolution and sensitivity. For high-quality imaging of microbial cells, we do not recommend pixel sizes greater than 9 µm.

6.2 Objectives

Under most circumstances, bacterial cells can be visualised with a $100\times$ objective. These are available in various apertures and which one to choose is dependent on the intended purpose. A higher numerical aperture (NA) will result in a trade-off with an increased amount of light (i.e. sensitivity) but decreased depth of field. Our current system utilises a $100\times$ NA 1.4 objective. Most standard $100\times$ objectives have an NA of 1.3 that is suitable for many applications.

To increase the magnification of the image, an Optivar lens (for Zeiss microscopes) can be used. While it will increase the size of the image without loss of resolution, the placing of the lens between the sample and the objectives will result in a significantly diminished signal to the CCD.

6.3 Filters

Filters for fluorescence consist of an excitation filter, a dichroic beam-splitter mirror and an emission filter. These filters are often bought as a set and are housed within a filter cube. Specialist filter suppliers include Omega Optical and Chroma Technology. It is essential to use the most appropriate, and optimal, filter set for the fluorescent protein or stain that is being used. Each protein has a specific excitation and emission wavelength (see Table 4.2). Both excitation and emission filters can let a narrow or wide range of wavelengths through. An emission filter with a wider bandwidth will be more sensitive but less specific. This can be an issue if dual fluorophores are being imaged in a single sample when fluorescence from one protein can be observed in the emission channel of the partner protein (crossover).

In order to visualise multiple fluorophores using dual or triple bandpass filter sets (e.g. GFP/mCherry), a filter wheel is recommended. The excitation filters are removed from the filter cube and placed in the filter wheel that can then be controlled to visualise the appropriate signal. When assembling a system, it is important that filter wheel and camera shutters have the appropriate drivers so that imaging can be performed rapidly and simply with a single click.

Neutral density filters (grey filters) reduce the transmission of light at all wavelengths by an equal amount. They are often used for samples that have unusually high fluorescence intensities (rare in bacteria) and to protect live samples from

Table 4.2 Fluorescence filter properties for fluorescent proteins and stains commonly used

Protein/stain	Excitation (nm)	Emission (nm)	Filter set (Chroma technology)
GFPmut1	488	507	41018
GFPmut3	501	511	41018
EYFP	514	527	41028
ECFP	439	476	31044v2
GFP/mCherry	488 587	507 610	59022
DAPI/FITC/Cy3	359 490 552	661 525 570	[a]61000v2SBX with a D395/40 excitation filter

[a]Current set available is 69000.

photodamage. Due to highly efficient optics and sensitive camera systems, the imaging of even low-abundance proteins is straightforward these days, and so neutral density filters should always be considered when conducting time-course experiments.

7 IMAGING SOFTWARE

The microscope and camera set-up need to be controlled by computer-driven imaging software. The available software can be used for either image acquisition or image processing. Software for acquisition is required to control such aspects as gain (detector sensitivity), binning and exposure time. Generally, only exposure time should need to be varied between samples. Binning increases the pixel size, so signal intensity can be increased, but at the expense of decreased resolution. MetaMorph® Software from Molecular Devices® (http://www.moleculardevices.com/Products/Software/Meta-Imaging-Series/MetaMorph.html) is a package that allows for both image acquisition and image processing for a variety of applications and is modular allowing simple customisation for specific requirements.

7.1 Image processing software

Image processing software can be specific either for microscopy (e.g. Imaris http://www.bitplane.com/) or for general image handling purposes (e.g. ImageJ http://rsbweb.nih.gov/ij/). Imaris is particularly useful when 3D or 4D imaging is required. ImageJ is freeware and a huge number of application-specific plugins have been developed that can be used for a vast array of microscopy purposes. All programmes permit image export in multiple file formats allowing the final figure preparation to be performed using programmes such as Adobe Photoshop®.

8 IMAGE PROCESSING

Image processing is one of the most important steps in localisation studies, ensuring the generation of high-quality images suitable for publication. After a focussed image is taken, the same field should be moved completely out of focus to allow a background image to be captured. This image can now be used to subtract any background fluorescence and autofluorescence that may be present, as well as being able to take into account micro-heterogeneities that can be present within the agarose pad. Using an out-of-focus image also takes into account the inherent variation in intensity across the field, which will vary from microscope to microscope. Figure 4.4 shows an in-focus image of RNAP-mCherry from *B. subtilis* (Panel A) and an out-of-focus image (Panel B). A linescan from the top left of this image to the bottom right is represented graphically in Panel C. It shows that the intensity of the fluorescence varies across the screen. This variation is also present in the focussed image shown in Panel A. By subtracting the out-of-focus image from

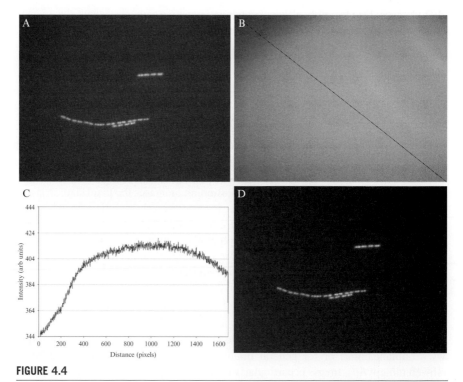

FIGURE 4.4

Image processing using background subtraction. The raw image is shown in (A). An out-of-focus image is shown in (B). A graphical representation of the linescan taken through B (diagonal line from top left to bottom right) is shown in (C), while the processed image is shown in (D).

the in-focus image, a processed image that is normalised for this intensity variation is created to give the smooth final image shown in Panel D. Although most image processing packages contain a 'flat field correction' function that is designed to take into account signal variation across an image, we have found that the above approach gives a more reliable and higher quality corrected image, particularly when imaging cells on agarose pads.

8.1 Signal comparison

Once the image has been processed, MetaMorph® has a very useful function that allows the intensity of fluorescence from labelled proteins in different strains to be compared directly. By creating a stack of two or more fields imaged under identical conditions, the intensity of fluorescence between the fields can be directly compared. This is done by firstly stacking the images to equalise their signal intensities and then presenting them as a montage. Figure 4.5 shows the results that equalising can achieve, using GFP-tagged metabolic enzymes CitC and CitZ from *B. subtilis*. When comparing the unequalised images in the top panels, the intensity and therefore the cellular concentration of the fluorescent protein fusions appear to be similar. However, by comparing the bottom two panels that were equalised for intensity, it is

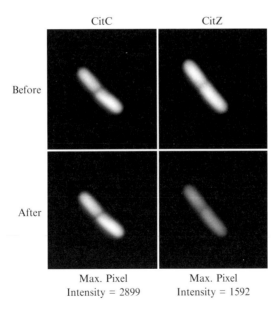

FIGURE 4.5

Equalising intensity for protein abundance estimation. The top panels show the processed images of CitC and CitZ before being equalised. The bottom two panels show the results of image stacking and montaging, which allow direct comparisons of fluorescent intensities to be observed.

apparent that CitC is more abundant than CitZ. This is consistent with the absolute cellular abundance for these two proteins under these conditions determined previously (Buescher et al., 2012).

9 CONCLUDING REMARKS

The development of fluorescent proteins as tools for imaging protein dynamics *in situ* has revolutionised our understanding of microbial cell biology. Today, many system packages are available from all the major microscope suppliers that enable high-quality imaging of fluorescently tagged microbial cells. For the more demanding user, bespoke systems can also be readily assembled. In this chapter, we have outlined and highlighted basic principles and approaches to constructing and imaging fluorescent protein fusions in some of the most common model microbes, and many of these approaches are transferable to other organisms. Once the basics have been mastered, there are now many systems available that enable more technically demanding studies such as protein–protein interaction (FRET/FLIM), 3D reconstruction, super-high resolution (e.g. DeltaVision OMX) and single-molecule tracking using total internal reflection microscopy.

Acknowledgements

This work was supported by grants from the Australian Research Council (ARC), National Health and Medical Research Council (NHMRC) and European Union.

References

Buescher, J. M., Liebermeister, W., Jules, M., Uhr, M., Muntel, J., Botella, E., Hessling, B., Kleijn, R. J., Le Chat, L., et al. (2012). Global network reorganization during dynamic adaptations of Bacillus subtilis metabolism. *Science* **335**, 1099–1103.

Chalfie, M., Tu, Y., Euskirchen, G., Ward, W. W., and Prasher, D. C. (1994). Green fluorescent protein as a marker for gene expression. *Science* **263**, 802–805.

Cormack, B. P., Valdivia, R. H., and Falkow, S. (1996). FACS-optimized mutants of the green fluorescent protein (GFP). *Gene* **173**, 33–38.

Datsenko, K. A. and Wanner, B. L. (2000). One-step inactivation of chromosomal genes in Escherichia coli K-12 using PCR products. *Proc. Natl. Acad. Sci. U.S.A.* **97**, 6640–6645.

Davies, K. M., Dedman, A. J., van Horck, S., and Lewis, P. J. (2005). The NusA:RNA polymerase ratio is increased at sites of rRNA synthesis in Bacillus subtilis. *Mol. Microbiol.* **57**, 366–379.

de Berardinis, V., Vallenet, D., Castelli, V., Besnard, M., Pinet, A., Cruaud, C., Samair, S., Lechaplais, C., Gyapay, G., et al. (2008). A complete collection of single-gene deletion mutants of Acinetobacter baylyi ADP1. *Mol. Syst. Biol.* **4**, 174.

Doherty, G. P., Fogg, M. J., Wilkinson, A. J., and Lewis, P. J. (2010). Small subunits of RNA polymerase: localization, levels and implications for core enzyme composition. *Microbiology* **156**, 3532–3543.

Fekete, R. A. and Chattoraj, D. K. (2005). A cis-acting sequence involved in chromosome segregation in Escherichia coli. *Mol. Microbiol.* **55**, 175–183.

Feucht, A. and Lewis, P. J. (2001). Improved plasmid vectors for the production of multiple fluorescent protein fusions in Bacillus subtilis. *Gene* **264**, 289–297.

Gordon, G. S., Sitnikov, D., Webb, C. D., Teleman, A., Straight, A., Losick, R., Murray, A. W., and Wright, A. (1997). Chromosome and low copy plasmid segregation in E. coli: visual evidence for distinct mechanisms. *Cell* **90**, 1113–1121.

Grkovic, S., Brown, M. H., Hardie, K. M., Firth, N., and Skurray, R. A. (2003). Stable low-copy-number Staphylococcus aureus shuttle vectors. *Microbiology* **149**, 785–794.

Hanahan, D. (1983). Studies on transformation of Escherichia coli with plasmids. *J. Mol. Biol.* **166**, 557–580.

Heim, R. and Tsien, R. Y. (1996). Engineering green fluorescent protein for improved brightness, longer wavelengths and fluorescence resonance energy transfer. *Curr. Biol.* **6**, 178–182.

Heim, R., Cubitt, A. B., and Tsien, R. Y. (1995). Improved green fluorescence. *Nature* **373**, 663–664.

Johnson, A. S., van Horck, S., and Lewis, P. J. (2004). Dynamic localization of membrane proteins in Bacillus subtilis. *Microbiology* **150**, 2815–2824.

Landgraf, D., Okumus, B., Chien, P., Baker, T. A., and Paulsson, J. (2012). Segregation of molecules at cell division reveals native protein localization. *Nat. Methods* **9**, 480–485.

Lau, I. F., Filipe, S. R., Soballe, B., Okstad, O. A., Barre, F. X., and Sherratt, D. J. (2003). Spatial and temporal organization of replicating Escherichia coli chromosomes. *Mol. Microbiol.* **49**, 731–743.

Lewis, P. J. and Marston, A. L. (1999). GFP vectors for controlled expression and dual labelling of protein fusions in Bacillus subtilis. *Gene* **227**, 101–110.

Lewis, P. J., Thaker, S. D., and Errington, J. (2000). Compartmentalization of transcription and translation in Bacillus subtilis. *EMBO J.* **19**, 710–718.

Liew, A. T., Theis, T., Jensen, S. O., Garcia-Lara, J., Foster, S. J., Firth, N., Lewis, P. J., and Harry, E. J. (2011). A simple plasmid-based system that allows rapid generation of tightly controlled gene expression in Staphylococcus aureus. *Microbiology* **157**, 666–676.

Meile, J. C., Wu, L. J., Ehrlich, S. D., Errington, J., and Noirot, P. (2006). Systematic localisation of proteins fused to the green fluorescent protein in Bacillus subtilis: identification of new proteins at the DNA replication factory. *Proteomics* **6**, 2135–2146.

Possoz, C., Filipe, S. R., Grainge, I., and Sherratt, D. J. (2006). Tracking of controlled Escherichia coli replication fork stalling and restart at repressor-bound DNA in vivo. *EMBO J.* **25**, 2596–2604.

Robinett, C. C., Straight, A., Li, G., Willhelm, C., Sudlow, G., Murray, A., and Belmont, A. S. (1996). In vivo localization of DNA sequences and visualization of large-scale chromatin organization using lac operator/repressor recognition. *J. Cell Biol.* **135**, 1685–1700.

Shu, X., Shaner, N. C., Yarbrough, C. A., Tsien, R. Y., and Remington, S. J. (2006). Novel chromophores and buried charges control color in mFruits. *Biochemistry* **45**, 9639–9647.

Viollier, P. H., Thanbichler, M., McGrath, P. T., West, L., Meewan, M., McAdams, H. H., and Shapiro, L. (2004). Rapid and sequential movement of individual chromosomal loci to specific subcellular locations during bacterial DNA replication. *Proc. Natl. Acad. Sci. U.S.A.* **101**, 9257–9262.

Webb, C. D., Graumann, P. L., Kahana, J. A., Teleman, A. A., Silver, P. A., and Losick, R. (1998). Use of time-lapse microscopy to visualize rapid movement of the replication origin region of the chromosome during the cell cycle in Bacillus subtilis. *Mol. Microbiol.* **28**, 883–892.

Werner, J. N., Chen, E. Y., Guberman, J. M., Zippilli, A. R., Irgon, J. J., and Gitai, Z. (2009). Quantitative genome-scale analysis of protein localization in an asymmetric bacterium. *Proc. Natl. Acad. Sci. U.S.A.* **106**, 7858–7863.

CHAPTER 5

Targeted and quantitative metabolomics in bacteria

Hannes Link*, Joerg Martin Buescher*,†, Uwe Sauer*,1

Institute of Molecular Systems Biology, ETH Zurich, Zurich, Switzerland
†*Biotechnology Research and Information Network AG, Zwingenberg, Germany*
1*Corresponding author. e-mail address: sauer@imsb.biol.ethz.ch*

1 INTRODUCTION

Metabolites are small non-polymer molecules with an atomic mass typically in the range of 50–1000 Da. The 1136 unique metabolites in the latest genome-scale metabolic model of *Escherichia coli* (Orth *et al.*, 2011) represent only a lower estimate for the number of metabolites that we can expect in a typical bacterium. In total, the KEGG database includes 5757 metabolites and other compounds in the metabolic pathways of different organisms (Hattori *et al.*, 2003). Experimentally, over 1500 distinct metabolite ions were detected in *E. coli* extracts (Fuhrer *et al.*, 2011).

Two different methodological aims can be addressed when investigating the metabolomic complexity of biological systems. Untargeted metabolic profiling aims to detect as many metabolites as possible in a sample and recent methods are able to detect a broad spectrum of 400–1500 metabolites (Madalinski *et al.*, 2008; Fuhrer *et al.*, 2011). Targeted metabolomics aims at the reliable and sensitive quantification of a pre-selected subset of metabolites, and mass spectrometric methods typically quantify absolute concentrations of around 100 metabolites (Bennett *et al.*, 2008; Buescher *et al.*, 2010). Here, we focus on targeted quantitative metabolomics in microorganisms, where the combination of gas- or liquid chromatography separation, coupled to mass spectrometry detection, is the most widely used analytical approach.

Data obtained with such quantitative methods reveal that the intracellular concentration of many metabolites is in the micro-molar concentration range and only few occur at a concentration greater than 10 mM. Depending on the environmental conditions, these dominating metabolites include glutamate, glutathione, fructose 1,6-bisphosphate and ATP that participate in many reactions (Bennett *et al.*, 2009; Buescher *et al.*, 2012). In various yeast strains, the total amino acid concentration represented up to 90% of the detected metabolites (Christen and Sauer, 2011). While amino acid concentrations remain rather stable, many low abundance metabolites exhibit a high degree of fluctuation. These fluctuations are a consequence of the flow equilibrium associated with the continuous supply and utilization of each metabolite in the metabolic network.

In sharp contrast to other 'omics' methods that assess mRNA or protein abundance, metabolomics provide a very fast and sensitive functional read-out that is directly linked to bacterial physiology. The metabolome can respond within a second or less to environmental changes (Kresnowati et al., 2006). This advantage comes, however, with the major technical challenge that extracting intracellular metabolites from bacteria is highly susceptible to errors and artefacts. Essentially, any change of the physiochemical milieu, such as temperature, oxygen supply or pH, during the sampling procedure can lead to measurable differences and can obscure the faithful representation of the intracellular metabolite composition.

Here, we describe our current set of methods to obtain representative snapshots of a specified set of metabolites in bacteria. Much more important than for other 'omics methods', the analysis requires robust and standardized cultivation conditions that maintain cellular homeostasis during sampling until metabolism is arrested. Our protocol is structured into four steps of the workflow, for each of which we provide several methods and comment on their pros and cons:

- *Cultivation* describes our standard protocols for (A) media, (B) inoculum and (C) cultivation. The method is optimized for a fast filtration sampling to obtain steady-state metabolite concentrations in bacteria.
- *Sampling* describes three methods of sampling for metabolome analysis: (A) fast filtration for liquid shake flask cultures with low biomass concentration, (B) whole cell broth extraction for sampling from cultures with high biomass concentration and (C) high-throughput sampling from 96-well deep well plates.
- *Simultaneous extraction and inactivation of metabolism* describe two extraction methods: (A) hot ethanol extraction for global metabolome analysis in one step and (B) acidic acetonitrile extraction for quantification of nucleotide triphosphates.
- *Analysis of metabolite extracts by a targeted mass spectrometry* describes our analytical method consisting of (A) separation by liquid chromatography, (B) detection by mass spectrometry and (C) peak integration. Further, we describe how to obtain quantitative data by using (D) U-^{13}C-labelled internal standards and (E) an estimation of the cytosolic volume to calculate intracellular concentrations.
- In the last section, we briefly discuss how the absolute metabolite concentrations, obtained as above, can be used to investigate the regulation of cellular metabolism and other biological functions.

2 CULTIVATION OF BACTERIA FOR METABOLOME ANALYSIS

The metabolomics standard initiative recommends best practice reporting standards to provide sufficient information about strains, inoculation procedure, cultivation conditions and sampling time points to allow meaningful comparison of data sets and interpretation (van der Werf et al., 2007b). These reporting standards are even more relevant for metabolomics than for other 'omics' technologies because intracellular metabolite concentrations are very sensitive to the precise physiological conditions and the growth phase of the culture. Here, we describe the standard cultivation protocol for quantitative comparison of steady-state metabolite concentrations in different strains and carbon sources that has been tested in several laboratories of the EU-funded BaSysBio consortium (Buescher et al., 2012, www.basysbio.eu).

2.1 Medium

The M9 minimal medium used for cultivation of *E. coli* is composed per litre of

- 200 mL of heat-sterilized base salt solution containing: 7.5 g L^{-1} $(NH_4)_2SO_4$, 2.5 g L^{-1} NaCl, 37.6 g L^{-1} $Na_2HPO_4 \cdot 2H_2O$ and 15.0 g L^{-1} KH_2PO_4;
- 700 mL heat-sterilized double-distilled water;
- 1 mL of heat-sterilized 1 M $MgSO_4$;
- 1 mL of heat-sterilized 0.1 M $CaCl_2$;
- 0.6 mL of filter-sterilized 0.1 M $FeCl_3 \cdot 6H_2O$;
- 2 mL of filter-sterilized 1 mM thiamine HCl (add only prior to growth experiment);
- 10 mL of a filter-sterilized trace element solution containing: 0.18 g L^{-1} $ZnSO_4 \cdot 7H_2O$, 0.12 g L^{-1} $CuCl_2 \cdot 2H_2O$, 0.12 g L^{-1} $MnSO_4 \cdot H_2O$ and 0.18 g L^{-1} $CoCl_2 \cdot 6H_2O$;
- 25 mL of the tested carbon source solution in water (heat or filter sterilized);
- Use heat-sterilized double-distilled water to fill up to 1 L.

The M9 minimal medium used for cultivation of *Bacillus subtilis* is composed per litre of

- 200 mL of heat-sterilized base salt solution containing: 42.5 g L^{-1} $Na_2HPO_4 \cdot 2H_2O$, 15 g L^{-1} KH_2PO_4, 5 g L^{-1} NH_4Cl, 2.5 g L^{-1} NaCl;
- 700 mL heat-sterilized double-distilled water;
- 1 mL heat-sterilized 0.1 M $CaCl_2$;
- 1 mL heat-sterilized 1 M $MgSO_4$;
- 1 mL filter-sterilized 50 mM $FeCl_3 \cdot 6H_2O$ + 100 mM citric acid;
- 10 mL filter-sterilized trace element solution containing: 0.17 g L^{-1} $ZnCl_2$, 0.1 g L^{-1} $MnCl_2 \cdot 4H_2O$, 0.06 g L^{-1} $CoCl_2 \cdot 6H_2O$, 0.06 g L^{-1} $Na_2MoO_4 \cdot 2H_2O$ and 0.043 g L^{-1} $CuCl_2 \cdot 2H_2O$;
- 25 mL of the tested carbon source solution in water (heat or filter sterilized);
- Use heat-sterilized double-distilled water to fill up to 1 L.

Note that the formation of precipitate after mixing all compounds is normal; therefore, shake the bottle well before withdrawing medium.

2.2 Inoculum

- Plate the tested strain on Luria Bertani (LB) agar plates.
- Transfer a colony to 5 mL of LB-medium in a 15-mL tube for pre-culturing and cultivate for 3–5 h until the culture has an OD_{600} of 0.5–1.
- Transfer 5 µL of the LB pre-culture to 5 mL of M9 medium in a 15-mL tube for a second pre-culturing step. Since pre-culturing is usually done overnight, it is convenient to prepare serial dilutions of the LB pre-culture to achieve an appropriate OD the next day.
- Add 35 mL of M9 medium to a 500-mL non-baffled shake flask and preheat for 15 min in an incubator at the desired temperature.
- Inoculate the shake flask with the M9 pre-culture. Ideally, the pre-culture should be in exponential growth phase with an OD_{600} value between 0.5 and 1.0.
- Determine the amount of pre-culture that has to be added for a starting OD_{600} of 0.05 in the 35 mL M9 medium of the shake flask. Pellet the cells in the required amount of pre-culture by centrifugation (3 min at $5000 \times g$, 37 °C). Discard the supernatant and re-suspend the cell pellet in 100 µL of fresh medium.
- Inoculate the shake flask with the re-suspended cells.

2.3 Cultivation

- Check the OD of the culture following inoculation by removing 1 mL and measuring the OD_{600}.
- Incubate the shake flask at the desired temperature on an orbital shaker with 5 cm shaking diameter at 300 rpm.
- Samples of supernatant (optional): after measuring the OD_{600}, transfer the undiluted sample into an Eppendorf-type tube and pellet the cells at $12,000 \times g$ for 3 min. Decant the supernatant and store it in a separate Eppendorf tube at -20 °C until required for the measurement of the concentrations of extracellular substrate(s) and by-products.
- Monitor the growth of the culture by sampling 1 mL of culture every hour and measuring the OD_{600}. Use appropriate dilutions in 9 g L^{-1} NaCl to ensure that the measured OD_{600} is in the linear range of the spectrophotometer (typically 0.02–0.3 OD units).
- Verify that the cells are growing exponentially ($OD_{600} < 1.5$) when taking samples for metabolome analysis. However, sampling at $OD_{600} < 0.5$ reduces the sensitivity of the analysis. The experiments should be performed in biological replicates, usually two to three independent shake flask cultivations. The reproducibility of physiological parameters, such as growth rate, substrate uptake rate and by-product formation, is a measure of stable and robust cultivation conditions. Reproducible physiology is a prerequisite for metabolomics experiments since they are highly sensitive to these parameters.

3 SAMPLING BACTERIAL CULTURES FOR METABOLOME ANALYSIS

Sampling for bacterial metabolome analysis is technically very challenging since all changes in the physiochemical milieu can instantaneously alter the metabolome. Generally, there are two options to obtain a realistic snapshot of intracellular metabolite concentrations, either the cells are sampled within less than 1 s or the cell's environment is maintained constant during sampling. Rapid sampling requires special devices that are designed to simultaneously withdraw the sample and arrest metabolism. Some examples are a sampling tube device (Weuster-Botz, 1997), an automated sampling device for pulse experiments (Schaefer et al., 1999), the stopped-flow sampling device (Buziol et al., 2002), a mini-plug flow reactor called BioScope (Visser et al., 2002), a single tube heat exchanger (Schaub et al., 2006) and a low pressure in situ sampling device (Hiller et al., 2007). Although such devices minimize the time between sampling and the arrest of metabolism, there is the problem of cell disintegration when using a cold quenching solution to arrest metabolism (usually aqueous methanol at −50 °C). This problem of cell damage and concomitant metabolite leakage was first observed for amino acids in *Corynebacterium glutamicum* and blamed on a general cold shock during quenching (Wittmann et al., 2004). Later, the authors showed that various other species of bacteria leak metabolites during cold methanol quenching (Bolten et al., 2007). Quenching solutions with protecting agents such as glycerol were used to reduce leakage; however, the loss of metabolites was still significant (Villas-Bôas and Bruheim, 2007; Link et al., 2008). Since the problem of metabolite leakage remains unsolved, we do not recommend cold methanol quenching. An alternative method is simultaneous quenching and extraction of the whole culture medium, referred to as whole cell broth extraction (Schaub et al., 2006; Bolten et al., 2007; Taymaz-Nikerel et al., 2009). A major drawback with whole cell broth extraction is the high level of contamination of the samples with salts from the medium, thus reducing the sensitivity and selectivity of LC–MS analysis. Furthermore, this method requires correction for metabolites in the extracellular milieu. Accurate metabolite quantification using whole cell broth extraction is possible at high biomass concentrations because of the more favourable ratio of intracellular to extracellular volume compared with low biomass concentration samples.

Since no generally accepted and applicable sampling method has emerged yet, we describe here a fast filtration protocol that produces minimal errors if applied in combination with the cultivation conditions described in Section 2. Sampling by quick filtration and subsequent washing has been discussed intensively. Bolten et al. (2007) consider fast filtration a suitable sampling technique if cells are washed with appropriate washing solutions. However, they restrict applicability of the method to metabolites with a relatively low turnover. Rabinowitz (2007) concludes that the method is excessively likely to cause artefacts due to alterations to the metabolome during filtering and washing. The success of this method critically depends on the washing solution. We acknowledge that successful application of this fast filtration method requires some training, but in our experience, it remains the method of choice for accurate metabolome sampling from bacteria in liquid culture of low to moderate biomass concentration. For cultivation with high cell density (>5 g$_{DryWeight}$ L^{-1}), we recommend extracting the whole culture broth as described in method B below.

The great advantage of fast filtration is the low contamination of the samples with salts, which dramatically improves the separation of sugar phosphates and also the sensitivity of the LC–MS. As a result, the number of quantifiable metabolites is much higher when compared to whole cell broth extraction technique. A further advantage is the complete removal of extracellular metabolites since the cells are continuously perfused with a washing medium. The method only requires minimal technical set-up as sampling is performed manually with a pipette.

Precaution. Shake flask culture cells should be recovered from the same phase of pseudo steady-state growth and the biomass concentration should be in a range of OD$_{600}$ 0.5–1 to prevent artefacts caused by oxygen limitation and the accumulation of by-products. In our experience, a fast-growing bacterial culture ($\mu = 0.6$ h^{-1}) of this biomass concentration depletes the oxygen dissolved in minimal medium within 1 min. The quantities of cells and volumes are given to detect metabolites with a concentration of 1–100 μmol L^{-1} in the LC–MS samples.

3.1 Fast filtration

Fast filtration was initially proposed by Wittmann *et al.* (2004), and we apply this method to obtain steady-state metabolite concentrations in bacteria under robust and stable growth conditions as described in Section 2 (Figure 5.1):

- Assemble a vacuum filter flask and preheat to the temperature used for cultivating cells in a thermostatically controlled room.

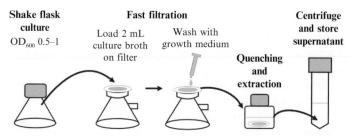

FIGURE 5.1

Schematic of fast filtration. Culture broth from a shake flask is transferred onto a nitrocellulose filter mounted on a vacuum device. Subsequently, cells are immediately washed with growth medium. Following filtration and washing, the intracellular metabolites are extracted directly from the cell-loaded filter disc and prepared for analysis. (For colour version of this figure, the reader is referred to the online version of this chapter.)

- Prior to sampling determine the OD and pH of the culture.
- Prepare 100 mL cultivation media to wash the cells, adjust to the pH of the culture at the time of sampling and preheat to the temperature used for cultivating the cells.
- Place a 0.45-μm pore size filter with 25 mm diameter (e.g. Millipore, no. HAWP02500) on the vacuum filter flask and apply a vacuum of ∼50 mbar.
- Pre-wash the filter with medium such that it is completely wetted.
- Transfer 2 mL of the culture broth onto the middle of the filter using a pipette. Load the filter carefully but be fast enough to prevent oxygen limitation in the pipette.
- Perfuse the filter with 5 mL of the preheated washing medium while ensuring that the filter never runs dry during loading and washing. The medium absorbed in the filter will provide nutrients only for a few seconds. Assure that the washing medium resembles conditions in the culture broth. Any modification of the washing medium can cause artefacts.
- After washing, immediately transfer the cell-loaded filter to the extraction solution using a pair of tweezers. The extraction solution will arrest metabolic activity and release metabolites from the cells (see Section 3).

3.2 Whole cell broth extraction

Fast filtration is not applicable if environmental conditions are likely to change during filtration, for example, when sampling from bioreactors with high-biomass concentrations and controlled substrate concentrations. In this case, we recommend extracting the whole culture broth and applying a differential method to quantify intracellular metabolites (Taymaz-Nikerel *et al.*, 2009). A major drawback is the contamination of the sample with salt from the medium, which causes reduced sensitivity and thus may reduce the number of metabolites that can be quantified. Furthermore, it is not possible to discriminate between the intracellular and extracellular location of a metabolite prior to extraction. However, in case of high biomass concentrations, the ratio of extracellular to intracellular volume is more favourable; therefore, the negative effects of extracellular medium in the sample are less critical.

Whole cell broth extraction
- Prepare 4 mL extraction solution in a centrifuge tube (see Section 3).
- Spray 1 mL of culture broth into the extraction solution. If possible use the bottom valve of the bioreactor or a rapid sampling device.
- The extraction solution will arrest any metabolic activity and release metabolites from the cells. At this point, extracellular metabolites can no longer be distinguished from intracellular metabolites.
- After extraction, centrifuge for 1 min at $12,000 \times g$ at 4 °C and store the supernatant at -80 °C.

Sampling of metabolites in the supernatant (optional)
- Prepare 4 mL extraction solution in a centrifuge tube (see Section 3).
- Transfer 2 mL of the culture broth into an Eppendorf-type tube and centrifuge at 4 °C.

- Transfer 1 mL of the supernatant into the extraction solution. The extraction solution will inactivate enzymes present in the culture supernatant.
- Centrifuge the extract for 1 min at $12,000 \times g$ and 4 °C and store the supernatant at −80 °C.

3.3 High-throughput sampling of bacterial cultures in 96-well format

Recently, rapid sampling was adapted to a 96-well format by using fritted plates for cultivation, which ensure fast sample transfer into cold methanol for quenching (Ewald et al., 2009). We adopted this method for bacteria using the whole cell broth extraction technique adapted for the 96-well format.

- Cultivate bacteria in a 96 deep well plate with fritted bottoms (Nunc, USA, no. 278011) using 500–600 μL medium and monitor growth by withdrawal of 20–25 μL samples.
- Determine OD_{600} just before sampling. All wells should be in a range of OD_{600} 0.5–1 for metabolome analysis.
- Fill each well of a 48-square deep well plate with 3.6 mL extraction solution and place it in a vacuum manifold.
- Apply a moderate vacuum of ∼50 mbar.
- Place the culture in the 96-well deep well plate with fritted bottoms on top of the vacuum manifold and transfer the culture to the extraction solution in the 48-well plate. Note that in this step two wells of the 96-well plate with ∼500 μL are pooled into a single well for extraction.
- Quickly remove the vacuum, take out the 48-well plate and add 100 μL internal standard (see Section **5.3**) to each well using a multi-channel pipette.
- Close the 48-well plate with a foil or rubber seal and incubate for 5 min at 80 °C in a water bath (use glass beads and manual shaking to increase extraction efficiency).
- After incubation, immediately cool the plate to ≤4 °C.
- Centrifuge the 48-well plate 10 min at 5000 rpm and 4 °C.
- Transfer the supernatant into a new 48-well plate and dry the extracts under vacuum at 0.12 mbar to complete dryness. Store the dry metabolite extracts at −80 °C until required.
- Re-suspend the dry metabolite extracts in 100 μL water using a multi-channel pipette and transfer into a 96-well storage plate.
- Centrifuge the storage plate 10 min at 5000 rpm and 4 °C and transfer 10–20 μL of the supernatant into another 96-storage plate which is then placed into the autosampler for analysis using a mass spectrometry method (optional: transfer supernatant into conical vials).

Using the above-described method will usually result in contamination of the sample with salts from the culture medium and with extracellular metabolites, both of which will interfere with the analysis. For specific questions, it might be applicable to wash and transfer the cultures to a low salt medium before sampling. This will reduce contamination of the samples with salt and remove a substantial amount of the extracellular metabolites that accumulate during cultivation. Such an approach, involving a rapid exchange of cultivation medium, was recently applied on a larger scale for the steady-state analysis of *E. coli* metabolism (Link *et al.*, 2009).

4 EXTRACTION OF METABOLITES FROM BACTERIA

The ideal extraction method should maximize the release of metabolites from the cell and at the same time minimize the degradation or conversion of these compounds. Extraction methods have involved combinations of freeze–thaw cycles, solvents, boiling and extreme pH. Six different methods for extracting metabolites from *E. coli* were compared by Maharjan and Ferenci (2003). They showed that the extraction methodology significantly influenced the results of metabolome analysis and, on the basis of their findings, recommended a cold methanol extraction method for global metabolome analysis. However, cold extraction does not always ensure the efficient deactivation of enzymes as reflected in the decomposition of nucleotide triphosphates when cold methanol/water was used (Kimball and Rabinowitz, 2006). A recent study reported that boiling ethanol is a suitable one-step method for global metabolome analysis, since the hot ethanol effectively extracts metabolites, denatures and precipitates proteins, is easily removed by evaporation, and has a minimal effect on pH (Winder *et al.*, 2008).

Here, we describe metabolite extraction involving hot ethanol/water (60:40) that ensures broad coverage of metabolites, yet is fast (3 min) and convenient. However, we also observed significant decomposition of nucleotide triphosphates in extracts prepared with hot ethanol due to exchange of phosphate groups among nucleotide phosphates. For this reason, we recommend acidic acetonitrile extraction to quantify high-energy metabolites such as adenosine triphosphates (Rabinowitz and Kimball, 2007).

4.1 Hot ethanol extraction of filtered samples

- Prepare 4 mL aliquots of 60:40 ethanol (p.a.)/water in small glass bottles and preheat to 78 °C in a water bath. Temperatures below the boiling point of ethanol facilitate the liquid handling.
- Transfer a nitrocellulose filter disc with sample into the extraction solution (see Section **2**). Ensure the filter is completely immersed in the hot ethanol.
- Add an internal standard (e.g. U-^{13}C-labelled metabolite extract).
- Incubate for 3 min at 78 °C, shaking vigorously every minute.
- Transfer the extract to a pre-cooled centrifuge tube on ice.
- Separate cell debris and nitrocellulose from metabolite extract by centrifugation at $12,000 \times g$ for 10 min at ≤ 4 °C.

- Dry the supernatants at 0.12 mbar to complete dryness in a vacuum concentrator. Dry metabolite extracts can be stored at $-80\ °C$ until analysis.
- Before analysis re-suspend dry metabolite extracts in 100 µL water.
- Centrifuge at $12,000 \times g$ for 10 min at $4\ °C$ and use 10–20 µL of the supernatant for mass spectrometry analysis.

4.2 Hot ethanol extraction of whole cell broth

- Prepare 4 mL 75:25 ethanol (p.a.)/water in 15 mL reaction tubes and preheat to $78\ °C$ in water bath (will yield a 60:40 ethanol:water ratio after addition of sample).
- Transfer 1 mL whole culture broth into extraction solution and mix by vigorous shaking for 3 s.
- Add internal standard (e.g. U-^{13}C-labelled metabolite extract).
- Incubate 3 min at $78\ °C$. If you observe sedimentation of cells, shake vigorously every minute.
- Separate cell debris and metabolite extract by centrifugation at $12,000 \times g$ for 10 min at $4\ °C$.
- Transfer supernatant to a fresh 15 mL tube.
- Add 1 mL 75:25 ethanol (p.a.)/water at $78\ °C$ to cell pellet, re-suspend and incubate at $78\ °C$ for 1 min.
- Separate cell debris and metabolite solution by centrifugation at $12,000 \times g$ for 10 min at $\leq 4\ °C$.
- Add this supernatant to the supernatant from the first extraction and dry at 0.12 mbar to complete dryness in a vacuum concentrator. Dry metabolite extracts can be stored at $-80\ °C$ until analysis.
- Before analysis, re-suspend dry metabolite extracts in 100 µL water. Depending on the medium composition, the sample might not dissolve completely.
- Centrifuge at $12,000 \times g$ for 10 min at $4\ °C$ and use 10–20 µL of the supernatant for mass spectrometry analysis.
- Carefully transfer 10 µL of clear supernatant to a vial or well plate used for analysis and dilute with 40–90 µL water (or mobile phase at start of LC gradient) to reduce salt concentration.

4.3 Acidic acetonitrile extraction of filtered samples

This method, adopted from that of the Rabinowitz group, is used to quantify nucleotide triphosphates that decompose during hot ethanol extraction (Kimball and Rabinowitz, 2006; Bennett et al., 2008).

- Prepare 4 mL of 40:40:20 acetonitrile/methanol/water with 0.1 M formic acid in small glass bottles and cool them to $-20\ °C$ in a freezer.
- Transfer a nitrocellulose filter disc with the sample into the extraction solution (see Section **2**). Ensure the filter is completely immersed in the cold acetonitrile solution.

- Add internal standard (e.g. U-^{13}C-labelled metabolite extract).
- Incubate for 15 min at $-20\ ^\circ$C in a freezer or a thermo bath.
- Add 400 µL of 15% NH_4OH to neutralize the acidic extract.
- Transfer the extract to a pre-cooled centrifuge tube.
- Separate cell debris and nitrocellulose from the metabolite extract by centrifugation at $12,000 \times g$ for 10 min at $\leq 4\ ^\circ$C.
- Dry the supernatants at 0.12 mbar to complete dryness in a vacuum concentrator. Dry metabolite extracts can be stored at $-80\ ^\circ$C until analysis.
- Before analysis re-suspend the dry metabolite extracts in 100 µL water centrifuge at $12,000 \times g$ for 10 min at $\leq 4\ ^\circ$C and use 10–20 µL of the supernatant for mass spectrometry.

5 ANALYSIS BY MASS SPECTROMETRY

As metabolism generally takes place in an aqueous environment (e.g. cytosol), most primary metabolites are hydrophilic and many posses ionic groups. Consequently, chromatographic separation can take place in the gas phase after derivatization or in liquid phase with normal phase, ion exchange, ion-pairing or hydrophilic interaction chromatography (Buescher et al., 2009). Note that metabolites containing pyrophosphate groups (e.g. nucleotide triphosphates, NAD) can not be analyzed by gas chromatography. Owing to the hydrophilic nature of most primary metabolites, electron spray ionization (ESI) is the most widely used interface from liquid chromatography to mass spectrometer. Of the various types of mass spectrometers that are commercially available, ion traps and time-of-flight instruments are the most versatile as they facilitate both quantification and compound identification by accurate mass determination. For targeted analysis, triple quadrupole instruments offer competitive sensitivity and a larger linear range of up to four orders of magnitude.

Due to the large differences in abundance and physicochemical properties of intracellular metabolites, no single published method can, to date, simultaneously quantify the >1000 metabolites presumed to be present in a microbial cell. For instance, in *E. coli* extracts, a combination of six methods was required for the quantification of 176 metabolites (van der Werf et al., 2007a,b) and a combination of two methods quantified 103 metabolites (Bennett et al., 2009). The ion-pairing LC triple quadrupole MS method described here is particularly suited for the targeted analysis of central carbon and energy metabolites (Buescher et al., 2010). It covers a wide range of metabolites (amino acids and precursors, sugar phosphates, organic acids, nucleotides, redox cofactors, coenzyme A esters) while still allowing for the separation of highly similar isomers such as sugar phosphates. Quantification was demonstrated for 138 compounds in pure standards; 76–104 of which could also be found in a diverse range of biological samples. As a prerequisite for this method, metabolites of interest must be water soluble.

5.1 Separation by ion-pairing chromatography

- Prepare mobile phase A by mixing 0.01 mol (1.85 g or 2.38 mL) tributylamine with 0.015 mol (0.90 g or 0.855 mL) acetic acid. Subsequently, add 50 mL methanol (LC–MS grade) and mix thoroughly. Then, fill to a total volume of 1 L with nanopure water of an electrical resistance greater than 16 MΩ. Unless working with an inline degasser, degas mobile phase A and mobile phase B (isopropanol).
- Connect the column (Waters Acquity T3, 150 mm × 2.1 mm × 1.8 μm, Waters Corporation, Milford, MA, USA) with pre-filter (KrudKatcher Ultra, Phenomenex, Torrance, CA, USA) and place in column oven set to 40 °C.
- Prior to first use, wash the column for 30 min with 100% mobile phase A, followed by a gradient to 100% mobile phase B over 30 min, then 30 min at 100% mobile phase B and finally, a gradient to 100% mobile phase A. Then, pre-condition the column with 10 injections of 10 μL of 10 mM potassium phosphate buffer (pH 7) to improve the peak shape of phosphorylated analytes.
- Equilibrate the column with mobile phase A, starting at a flow rate of 0.15 mL min^{-1}. Slowly increase the flow rate to 0.4 mL min^{-1}. Take care to keep the back pressure below 1000 bar.
- Connect the ultrahigh performance liquid chromatography system to the ESI source.
- Check the backpressure to determine if the liquid path (in particular, the column and the pre-filter) is clean; typical values for a clean system are between 600 and 750 bar.
- Check the mass spectrometry signal at 60 m/z (the M+1 isotopologue of acetate in negative mode) to ensure proper operation of the ESI source. The signal should be stable and within 10% of the value obtained directly after optimizing the settings of the ESI source.
- Separation is obtained by a combined mobile phase and flow rate gradient.

Time (min)	B (%)	Flow (mL min^{-1})
0	0	0.4
5	0	0.4
10	2	0.4
11	9	0.35
16	9	0.25
18	25	0.25
19	50	0.15
25	50	0.15
26	0	0.15
32	0	0.4
36	0	0.4

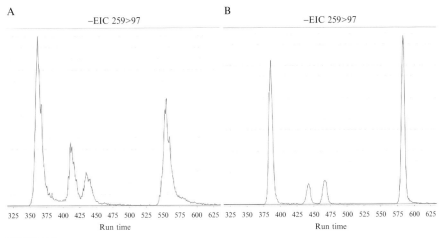

FIGURE 5.2

The separation of hexose phosphate isomers (from left to right: glucose-6-phosphate, fructose-6-phosphate, glucose-1-phosphate, fructose-1-phosphate) on a used column (A) is severely reduced compared to a new column (B). (For colour version of this figure, the reader is referred to the online version of this chapter.)

- The width of the chromatographic peaks with this gradient can be as short as 10 s. Thus, a data acquisition rate of more than 2 scans/(s compound) is required for reliable peak integration. If using an MS/MS-type mass spectrometer, the gradient may need to be split into several time segments, depending on the required time per scan and the number of target metabolites.
- Check the chromatographic separation by injecting 10 µL of a 5-µM standard mixture. The most sensitive test is the separation and peak shape of sugar-phosphate isomers (Figure 5.2).
- Prepare samples in water
  - Low concentrations of organic solvents can be tolerated, that is, <5% methanol and <1% other organic solvents. Higher organic solvent concentrations will interfere with the retention and separation of analytes. If analytes in the standard mixture have a concentration of 100 µM in 50% methanol, a 1:20 dilution of this mixture can be used straight away.
  - High salt concentrations can affect separation and peak shapes.
  - Typically, dissolve a dried extract obtained from a 1-mL culture of OD_{600} 1.0 ($\sim$50 $mg_{DryWeight}$) in 100 µL water.
- Typical injection volumes are between 2 and 20 µL.

5.2 Detection by ESI-MS/MS

- Prior to first use, optimize the ESI parameters. To this end, provide a continuous flow of 50 µL min^{-1} of 100 µM of a relevant metabolite

(e.g. glucose-6-phosphate) in 25:75 methanol:water to the ESI source using a syringe pump. Set slow-changing parameters (spray voltage and all gas flows and temperatures) to medium values. Then, optimize these parameters one after the other to improve signal intensity and the signal-to-noise ratio.
- For each metabolite of interest, optimize fast-changing parameters (tube lens voltage, collision energy, fragment ion mass). Provide a continuous flow of 20 µL min^{-1} of a 100 µM solution in 50:50 methanol water to the ESI source using a syringe pump. Most manufacturers provide automated software routines for the optimization of the compound-specific parameters. If possible, record the optimized tube lens voltage and collision energy for multiple fragments.
- Generally, the most intense fragment will provide the most sensitive quantification. However, in some cases, the use of alternative fragments can prevent interference with metabolites of similar retention time. For instance, galactose-1-phosphate and glucose-1-phosphate both yield a 241-m/z fragment ion, while fructose-6-phosphate and mannose-1-phosphate, which elute close by, do not, even though all four compounds are structural isomers.
- Ion suppression is a well-known source of error in ESI-MS. However, U-^{13}C-labelled internal standards can compensate this potential problem.
- With this method, metabolites can elute in a time window of 10–60 s. Reliable peak integration requires at least 20 data points per scan. Depending on the number of target metabolites, it might be necessary to not scan for all metabolites during the entire duration of the gradient but rather to divide the gradient into time segments. Thus, in one segment, only the metabolites that are eluted are quantified.
- In the long run, regular cleaning and maintenance is essential for metabolite quantification. In our experience, the ion entry (orifice or ion transfer capillary) should be cleaned weekly. Monthly cleaning of the Q00, skimmer, tube lens and, depending on the device, Q0 can be performed by an experienced user.
- In positive mode, tributylamine causes a strong and persistent signal at 186 m/z that can interfere with the tyrosine signal regularly used for mass axis calibration. If available, using a different ESI source for mass axis calibration significantly alleviates this problem. Otherwise, intensive cleaning is required before the mass spectrometer can be used in positive mode.

5.3 Peak integration of LC–MS/MS data

When working with manufacturer-provided software, manual peak integration might be necessary for analytes with two or more phosphate groups and for groups of isomers.

- Organic acids (e.g. fumaric acid) can suffer from interference with background signal that can hamper automatic base line detection.
- If unexpected or non-Gaussian peak shapes occur, check the signal of the corresponding U-^{13}C-labelled analyte. If the retention time and peak shape match, the peak can usually be used.

- The traces of some target metabolites can contain additional peaks of unknown metabolites. As retention times are stable within a few seconds for consecutive injections of samples of similar composition, the correct peak can easily be identified by comparing the retention time to that in a chemically pure standard.

5.4 Isotope ratio-based quantification of intracellular metabolite concentrations

Several isotope ratio-based approaches have been proposed for metabolome analysis (Mashego et al., 2004; Bennett et al., 2008). The general principle is that the amount of isotopically labelled internal standard and analyte is quantified separately using MS-based metabolomics and the ratio of both measurements is used for all further calculations. The internal standard is spiked into the solution of the extract right after sampling and undergoes similar degradation and loss during sample preparation. Isotopologues are ideal internal standards in MS-based metabolomics since standard and analyte have the same physicochemical properties but different masses. U-^{13}C isotopologues of all metabolites are relatively easily obtained from cells grown on minimal medium with U-^{13}C glucose as the sole carbon source. Usually, U-^{13}C-labelled cell extracts are obtained from yeast cultivated in a fed-batch fermentation on 100% U-^{13}C-labelled substrates (Wu et al., 2005). Here, we first describe a yeast fermentation to obtain large quantities of internal standard within 3 days of work. The second method describes a quick anaerobic cultivation to obtain labelled cell extract within 1 day of work.

1. **Batch fermentation of yeast on U-^{13}C-labelled glucose**
   This method describes fermentation in a bioreactor to obtain 1 L of U-^{13}C-labelled yeast culture.

   U-^{13}C glucose minimal medium
   The minimal medium used for the fed-batch cultivation of yeast is composed per litre of
   - 200 mL of a heat-sterilized base salt solution containing: 25 g L^{-1} (NH$_4$)$_2$SO$_4$, 2.5 g L^{-1} MgSO$_4$·7H$_2$O and 15.0 g L^{-1} KH$_2$PO$_4$.
   - 751 mL filter-sterilized KH–phthalate–water (10 mM, pH 5).
   - 1 mL of heat-sterilized vitamin solution containing 50 mg L^{-1} biotin, 1 g L^{-1} Ca-pantothenate, 1 g L^{-1} nicotinic acid, 25 g L^{-1} inositol, 1 g L^{-1} pyridoxine, 200 mg L^{-1} p-aminobenzoic acid, 1 g L^{-1} thiamine.
   - 10 mL of a filter-sterilized trace element solution containing: 0.15 g L^{-1} EDTA, 0.45 g L^{-1} ZnSO$_4$·7H$_2$O, 0.03 g L^{-1} CoCl$_2$·6H$_2$O, 0.10 g L^{-1} MnCl$_2$·4H$_2$O, 0.03 g L^{-1} CuSO$_4$·5H$_2$O, 0.45 g CaCl$_2$·2H$_2$O, 0.30 g L^{-1} FeSO$_4$·7H$_2$O, 0.04 g L^{-1} NaMoO$_4$·2H$_2$O, 0.10 g L^{-1} H$_3$BO$_3$, 0.01 g L^{-1} KI.
   - 40 mL of filter-sterilized U-^{13}C glucose solution (250 g L^{-1}).

   Bioreactor set-up
   - 2.1 L bioreactor with two Rushton turbines (e.g. Bioengineering KLF).

- 1 sampling port in the bottom.
- Off-gas cooling at 3 °C and off-gas analysis.
- Air is supplied at 1 vvm. In order to remove CO_2, the air is sparged through a 2-M KOH solution.
- Cultivations conditions: 1000 rpm stirrer speed, 30 °C temperature and pH 5 controlled with addition of 2 M HCl and 2 M NaOH, no overpressure.

Inoculum
- Inoculate 5 mL normal glucose minimal medium with yeast and grow over night at 30 °C. *Saccharomyces cerevisiae* CEN.PK, *S. cerevisiae* FY4 and *Pichia pastoris* have all been shown to be suitable (Kümmel *et al.*, 2010; Carnicer *et al.*, 2011).
- Add 35 mL of U-^{13}C glucose medium to a 500 mL non-baffled shake flask and preheat 15 min in the incubator at 30 °C.
- Inoculate the shake flask with the overnight pre-culture. Ideally, the pre-culture should be in exponential growth phase and has an OD_{600} of 1.0–1.5.

Fermentation
- Inoculate 965 mL U-^{13}C glucose medium in the bioreactor with the 35 mL of the exponentially growing shake flask pre-culture to obtain a starting OD_{600} of about 0.05.
- After about 17 h, the late ethanol-growth phase is reached (verify by growth rate and shift in the respiratory quotient, i.e. the ratio of O_2 consumption to CO_2 production).
- Sample 200 mL as described below.
- Add 2.5 g U-^{13}C glucose and continue cultivation for 5 min.
- Sample the rest of the culture in lots of 200 mL as described below.

Sampling and quenching procedure
- Prepare 5 L quenching solution: 60:40 methanol/water with 10 mM ammonium acetate. Adjust pH to pH 7.5 with 25% ammonium hydroxide.
- Distribute the quenching solution in 800 mL portions over six 1-L bottles and cool to −40 °C using a cryostat.
- Assemble a quenching device: half fill an ice bucket or a large beaker with denatured ethanol (methylated spirits) and add about 1.5 kg dry ice. Place the bucket on a magnetic stirrer and then place a 2-L measuring cylinder with a large magnetic stirrer bar in the bucket. Make sure the level of the ethanol in the ice bucket reaches up to the 800 mL level of the cylinder. Switch on the stirrer and pre-cool for 15 min.
- Just before sampling, add 800 mL quenching solution (at −40 °C) in the cylinder. Sample 200 mL of broth directly into the quenching solution while stirring continuously.
- Transfer the quenched suspension into two 1-L centrifuge buckets and centrifuge at $4000 \times g$ and −20 °C until supernatant is clear.

- Decant and discard the supernatant.
- Immediately, add 500 mL quenched yeast suspension, prepared as described above while the centrifuge is running, on top of each pellet. Repeat until the entire content of the bioreactor has been harvested.
- After the last aliquot of quenched yeast suspension has been centrifuged, re-suspend each pellet in 200 mL of fresh quenching solution. Aliquot 20 mL into 50-mL Falcon tubes. Keep all components below −20 °C throughout the entire procedure.
- Centrifuge again and discard the supernatant.
- Store the biomass at −80 °C until extraction.

Extraction
- Prepare 1.1 L extraction solution: 75:25 ethanol/water with 10 mM ammonium acetate pH 7.0.
- Aliquot 24 × 20 mL of extraction solution in 50-mL Falcon tubes and preheat to 78 °C.
- Pre-warm the quenched samples to −20 °C in a cryostat.
- Pour 20 mL preheated extraction solution directly from the Falcon tube to the pellet of the quenched sample and re-suspend immediately using a 25-mL pipette.
- Incubate the sample in a water bath for 3 min at 78 °C, vortexing every 60 s.
- Incubate the extracts in the cryostat at −20 °C until centrifugation.
- Separate cell debris from metabolite extract by centrifugation at 4000 × g and <4 °C.
- Pool all of the extracts and store at −80 °C as aliquots.

2. **Anaerobic batch cultivation of *E. coli* on 100% U-^{13}C-labelled glucose**
   This method describes a convenient bench top cultivation to obtain internal standard from an anaerobic culture of *E. coli*. Anaerobic conditions are required to prevent the incorporation of ^{12}C carbon in metabolites via CO_2 assimilation. However, it should be noted that the anaerobic conditions significantly reduce the biomass yield and thus the amount of U-^{13}C-labelled metabolites.
   - Prepare 100 mL of a minimal salt media with 2 g L^{-1} U-^{13}C glucose and degas with nitrogen.
   - Grow an anaerobic overnight pre-culture in 5 mL media to an OD_{600} of about 1.
   - Prepare an anaerobic flask and fill with 95 mL of medium.
   - Inoculate the anaerobic flask with 5 mL of the pre-culture.
   - After ∼6–7 h, when the culture has reached an OD_{600} of 0.5, start sampling.
   - Prepare five glass bottles filled with 10 mL of 60:40 ethanol/water and preheat to 80 °C.
   - Place a 0.45-μm pore size nitrocellulose filter with 90 mm diameter on a vacuum filter flask and apply a moderate vacuum.

- Transfer 20 mL of the culture broth ($OD_{600} \sim 0.5$) onto the middle of the filter using a pipette. Try to load the filter evenly with cells and ensure that the filter never runs dry during loading.
- Wash the recovered cells with a fourfold diluted salt's medium containing 1 g L^{-1} U-^{13}C glucose—the diluted medium reduces the contamination of the internal standard with salts.
- Transfer the cell-loaded filters very quickly to the glass bottles with hot ethanol using a pair of tweezers.
- Incubate 3 min at 78 °C.
- Transfer the extract to a pre-cooled centrifuge tube on ice.
- Separate cell debris and nitrocellulose from metabolite extract by centrifugation at 12,000 × g for 10 min at ≤ 4 °C.
- Store supernatant at −20 °C.
- Repeat four times until the whole culture has been harvested (optional: washing a fraction without carbon source will increase some low abundant metabolites like cyclic AMP or phosphoenolpyruvate).
- Vacuum dry the extract and re-suspend in 1 mL water.
- Pool all cell extracts and remove proteins by ultra-filtration.

3. Spiking with U-^{13}C-labelled standards and evaluation of results

- Add 100 μL of U-^{13}C-labelled internal standard to the extraction solution. Internal standard and sample are added at the same time to the extraction solution (see Section 4).
- Calculate the ratio $\emptyset_i$ of the signal of unlabelled metabolite i and the signal of the corresponding U-^{13}C-labelled metabolite by mass spectrometry. This ratio equals the moles of metabolite extracted from the cells (n_i^{cell}) and moles of U-^{13}C-labelled metabolite from the internal standard (n_i^{IS}):

$$\emptyset_i = \frac{n_i^{cell}}{n_i^{IS}} = \frac{c_i^{cell} V^{cell}}{c_i^{IS} V^{IS}}$$

where V^{cell} is the intracellular volume of the extracted biomass and V^{IS} is the volume of internal standard added during extraction (here 100 μL). In order to calculate the intracellular metabolite concentration c_i^{cell} from this ratio, the concentration of the internal standard (c_i^{IS}) is determined with a metabolite standard mixture, which is prepared by dissolving the pure chemicals separately.

4. Metabolite standard mixture

- Prepare stock solutions of the pure metabolites separately, if possible, at a concentration of 100 mM in water. For acids, neutralize the pH with ammonium. To this end, weigh the pure compounds into a 1.5-mL Eppendorf-type tube using an analytical balance. Use the calculated amount just as a reference and record the exact amount. Then, calculate the final volume based on this amount and add water or buffer accordingly.
- Mix single compound standards such that every compound has a concentration of 100 μmol L^{-1} in the final volume. Add 10 mM ammonium acetate pH 7.2 and

methanol such that the final stock has 50:50 methanol/buffer. The advantage of adding methanol is that the mixture remains liquid at $-20\,°C$ thus facilitating subsequent handling. Aliquot calibration mixture and store at $-80\,°C$ until analysis.
- When working with extremely thermo-instable compounds, such as redox cofactors, exempt them from the calibration mixture. Rather, prepare fresh standards directly before analysis.

For calibration, a constant amount of U-^{13}C internal standard is mixed with a dilution series of the calibration mixture. The range of concentrations in the dilution series should ideally cover the whole range of concentrations that occurs in the cell extract.

- Prepare seven levels of metabolite standard by a 1:3 dilution series of the metabolite calibration mixture. The resulting concentrations are 100, 33.3, 11.1, 3.7, 1.23, 0.41 and 0.14 µM.
- Mix 15 µL of each metabolite standard with 15 µL of the internal standard.
- Vacuum dry the standards and re-suspend in 15 µL pure water.
- Calculate the ratio $\emptyset_i$ of the signal of ^{12}C metabolites and the signal of the corresponding U-^{13}C-labelled metabolites. This ratio equals the moles of ^{12}C metabolite in the calibration level j ($n_i^{cal_j}$) and U-^{13}C metabolite from the internal standard (n_i^{IS}).

$$\emptyset_i = \frac{n_i^{cal_j}}{n_i^{IS}} = \frac{c_i^{cal_j} V^{cal}}{c_i^{IS} V^{IS}}$$

where $c_i^{cal_j}$ and V^{cal} are the concentration and the volume of the calibration level j added (here 15 µL) and V^{IS} is the volume of the internal standard added (here 15 µL). The unknown concentration in the internal standard c_i^{IS} follows from:

$$c_i^{IS} = \frac{c_i^{cal_j} V_i^{cal}}{\emptyset_i V_i^{IS}}$$

Since the ratio is determined at different calibration levels $c_i^{cal_j}$, use a regression analysis to get the best estimate for c_i^{IS}. Usually, a linear regression of ratio versus concentration in the calibration level will work.
- Using the thereby estimated concentration of the internal standard c_i^{IS}, the number of moles of metabolite i extracted from bacteria follows from:

$$n_i^{cell} = \emptyset_i n_i^{IS} = \emptyset_i c_i^{IS} V_{IS}$$

5.5 Normalizing quantified metabolites to biomass or total cell volume

In order to obtain quantitative data, the ratio $\emptyset_i$ or the extracted moles n_i, which have been determined as described in Section 5.4, have to be normalized. Usually, the amount of extracted biomass m_x is given as mL OD or $mg_{DryWeight}$. For comparison of relative changes, it is sufficient to normalize the ratio to the amount of biomass ($\emptyset_i/m_x$) assuming the cell morphology remained unchanged during the experiment.

In order to obtain absolute numbers for the intracellular concentrations, the cytosolic volume of the biomass is required. In general, the intracellular metabolite concentration as follows:

$$c_i^{cell} = \frac{n_i^{cell}}{V^{cell}} = \frac{n_i}{V_{Sample} OD_{Sample} v_{spec}}$$

where V_{Sample} is the total volume of extracted culture and OD_{sample} is the OD_{600} at the time of sampling. The specific cell volume v_{spec} (in $\mu L\ L^{-1} OD^{-1}$ or $\mu L\ mg_{DryWeight}^{-1}$) should be determined for the investigated strain and cultivation conditions. Recently, it was shown that 1 mL sample of an *E. coli* culture with an OD_{600} of 1 has a total cell volume of 2.4–4.9 μL depending on cultivation conditions (Volkmer and Heinemann, 2011).

6 INTERPRETATION OF METABOLITE DATA

The possibilities for interpretation of metabolomics data very much depend on the level of quantification that was achieved on the analytical side (Table 5.1). The metabolic workflow described above is suitable for the generation of quantitative data on absolute concentrations. If the calibration of the method to external standards or the conversion to intracellular concentrations based on cell volume is omitted, the generated data can perfectly well be used as relative-quantitative data. If U-^{13}C-labelled internal standard is omitted, the data can still be used as semi-quantitative data.

Table 5.1 Four Levels of Quantification in Metabolomics

	Typical applications	Typical questions
Non-quantitative	Compound identification	Is metabolite A present in the sample?
Semi-quantitative	Very high-throughput screening	Is the concentration of metabolite A in sample X higher, lower or similar compared to sample Y
		Which metabolites exhibit the greatest difference in concentration between samples X and Y
Relative-quantitative	Metabolomics as input to statistical analyses	Does the concentration of metabolite A correlate with environmental parameter X? Can samples be classified into groups?
Absolute-quantitative	Integration with other absolute data, kinetic modelling, thermodynamic modelling	How does the concentration of metabolite A relate to the K_m of enzyme X? What thermodynamic constraint does the concentration of metabolite A impose on reaction Y?

Absolute-quantitative data are a prerequisite for all thermodynamic analysis and for estimations of *in vivo* kinetic parameters such as binding and dissociations constants. The thermodynamic driving force constrains the direction of metabolic flux and is a function of absolute metabolite concentrations. Thermodynamic analysis provides a feasibility and consistency check of the data for metabolic fluxes and metabolite concentrations and reveals sites of active metabolite regulation (Kümmel *et al.*, 2006). Further, absolute metabolite concentrations facilitate the identification of *in vivo* kinetic properties of catalytic enzymes, by dynamic models that sum up kinetic rate laws in differential equations. A mechanistic kinetic model was used to infer *in vivo* kinetic parameters of central carbon metabolism of *E. coli* (Chassagnole *et al.*, 2002). Other models could additionally describe the function of metabolic modules, such as those used for the kinetic models of sphingolipid metabolism in yeast (Alvarez-Vasquez *et al.*, 2005) and ammonia assimilation in *E. coli* (Yuan *et al.*, 2009). The scope of future studies is to dissect the different levels of metabolite regulation that can occur as a result of the occupation of the active site (Bennett *et al.*, 2009; Fendt *et al.*, 2010) or of various allosteric metabolite–protein interactions (Gerosa and Sauer, 2011).

7 OUTLOOK

In the upcoming years, we anticipate that an increasing number of laboratories will engage in metabolomics. The commercial availability of sampling devices, decreasing prices of analytical devices of sufficient sensitivity and availability of easy-to-use software solutions will further facilitate this process. To assist existing newcomers to the field, in this compilation of the current state-of-the-art methods, we have focused on completeness and on the practical aspects for the various stages of a metabolomics experiment.

The amount of metabolomics data will increase not only in the number of experiments performed but also in the amount of data generated per experiment. Improvements in sensitivity of mass spectrometers will provide access to metabolites that are below the current limit of quantification. Furthermore, ultrahigh-throughput analysis dramatically increases the number of samples that can be measured (Fuhrer *et al.*, 2011). First steps have already been taken to add another dimension of complexity by quantifying metabolites with single-cell resolution (Amantonico *et al.*, 2010; Heinemann and Zenobi, 2011).

The increasing amount and depth of metabolomics data will intensify the need for standardized data handling and centralized data storage, similar to what is already available for genome sequences and transcriptomics. The combined study of multiple levels of large-scale data such as metabolomics, proteomics and transcriptomics then enables biological insights that are not be accessible through the study of any one of these levels alone (Fendt *et al.*, 2010; Buescher *et al.*, 2012).

References

Alvarez-Vasquez, F., et al. (2005). Simulation and validation of modelled sphingolipid metabolism in Saccharomyces cerevisiae. *Nature* **433**, 425–430.

Amantonico, A., Urban, P. L., and Zenobi, R. (2010). Analytical techniques for single-cell metabolomics: state of the art and trends. *Anal. Bioanal. Chem.* **398**, 2493–2504.

Bennett, B. D., Yuan, J., Kimball, E. H., and Rabinowitz, J. D. (2008). Absolute quantitation of intracellular metabolite concentrations by an isotope ratio-based approach. *Nat. Protoc.* **3**, 1299–1311.

Bennett, B. D., Kimball, E. H., Gao, M., Osterhout, R., Van Dien, S. J., and Rabinowitz, J. D. (2009). Absolute metabolite concentrations and implied enzyme active site occupancy in Escherichia coli. *Nat. Chem. Biol.* **5**, 593–599.

Bolten, C. J., Kiefer, P., Letisse, F., Portais, J.-C., and Wittmann, C. (2007). Sampling for metabolome analysis of microorganisms. *Anal. Chem.* **79**, 3843–3849.

Buescher, J. M., Czernik, D., Ewald, J. C., Sauer, U., and Zamboni, N. (2009). Cross-platform comparison of methods for quantitative metabolomics of primary metabolism. *Anal. Chem.* **81**, 2135–2143.

Buescher, J. M., Moco, S., Sauer, U., and Zamboni, N. (2010). Ultrahigh performance liquid chromatography–tandem mass spectrometry method for fast and robust quantification of anionic and aromatic metabolites. *Anal. Chem.* **82**, 4403–4412.

Buescher, J. M., Liebermeister, W., Jules, M., et al. (2012). Global network reorganization during dynamic adaptations of Bacillus subtilis metabolism. *Science* **335**, 1099–1103.

Buziol, S., Bashir, I., Baumeister, A., Claassen, W., Noisommit-Rizzi, N., Mailinger, W., and Reuss, M. (2002). New bioreactor-coupled rapid stopped-flow sampling technique for measurements of metabolite dynamics on a subsecond time scale. *Biotechnol. Bioeng.* **80**, 632–636.

Carnicer, M., Canelas, A. B., Ten Pierick, A., Zeng, Z., van Dam, J., Albiol, J., Ferrer, P., Heijnen, J. J., and van Gulik, W. (2011). Development of quantitative metabolomics for Pichia pastoris. *Metabolomics* **8**, 284–298.

Chassagnole, C., Noisommit-Rizzi, N., Schmid, J. W., Mauch, K., and Reuss, M. (2002). Dynamic modeling of the central carbon metabolism of Escherichia coli. *Biotechnol. Bioeng.* **79**, 53–73.

Christen, S. and Sauer, U. (2011). Intracellular characterization of aerobic glucose metabolism in seven yeast species by 13C flux analysis and metabolomics. *FEMS Yeast Res.* **11**, 263–272.

Ewald, J. C., Heux, S., and Zamboni, N. (2009). High-throughput quantitative metabolomics: workflow for cultivation, quenching, and analysis of yeast in a multiwell format. *Anal. Chem.* **81**, 3623–3629.

Fendt, S.-M., Buescher, J. M., Rudroff, F., Picotti, P., Zamboni, N., and Sauer, U. (2010). Tradeoff between enzyme and metabolite efficiency maintains metabolic homeostasis upon perturbations in enzyme capacity. *Mol. Syst. Biol.* **6**, 356.

Fuhrer, T., Heer, D., Begemann, B., and Zamboni, N. (2011). High-throughput, accurate mass metabolome profiling of cellular extracts by flow injection-time-of-flight mass spectrometry. *Anal. Chem.* **83**, 7074–7080.

Gerosa, L. and Sauer, U. (2011). Regulation and control of metabolic fluxes in microbes. *Curr. Opin. Biotechnol.* **22**, 566–575.

Hattori, M., Okuno, Y., Goto, S., and Kanehisa, M. (2003). Development of a chemical structure comparison method for integrated analysis of chemical and genomic information in the metabolic pathways. *J. Am. Chem. Soc.* **125**, 11853–11865.

Heinemann, M. and Zenobi, R. (2011). Single cell metabolomics. *Curr. Opin. Biotechnol.* **22**, 26–31.

Hiller, J., Franco-Lara, E., Papaioannou, V., and Weuster-Botz, D. (2007). Fast sampling and quenching procedures for microbial metabolic profiling. *Biotechnol. Lett.* **29**, 1161–1167. doi:http://dx.doi.org/10.1007/s10529-007-9383-9.

Kimball, E. and Rabinowitz, J. D. (2006). Identifying decomposition products in extracts of cellular metabolites. *Anal. Biochem.* **358**, 273–280.

Kresnowati, M. T., van Winden, W. A., Almering, M. J., ten Pierick, A., Ras, C., Knijnenburg, T. A., Daran-Lapujade, P., Pronk, J. T., Heijnen, J. J., and Daran, J. M. (2006). When transcriptome meets metabolome: fast cellular responses of yeast to sudden relief of glucose limitation. *Mol. Syst. Biol.* **2**, 49.

Kümmel, A., Panke, S., and Heinemann, M. (2006). Putative regulatory sites unraveled by network-embedded thermodynamic analysis of metabolome data. *Mol. Syst. Biol.* **2**, 34.

Kümmel, A., Ewald, J. C., Fendt, S. M., Jol, S. J., Picotti, P., Aebersold, R., Sauer, U., Zamboni, N., and Heinemann, M. (2010). Differential glucose repression in common yeast strains in response to HXK2 deletion. *FEMS Yeast Res.* **10**, 322–332.

Link, H., Anselment, B., and Weuster-Botz, D. (2008). Leakage of adenylates during cold methanol/glycerol quenching of Escherichia coli. *Metabolomics* **4**, 240–247.

Link, H., Anselment, B., and Weuster-Botz, D. (2009). Rapid media transition: an experimental approach for steady state analysis of metabolic pathways. *Biotechnol. Prog.* **26**, 1–10.

Madalinski, G., Godat, E., Alves, S., Lesage, D., Genin, E., Levi, P., Labarre, J., Tabet, J. C., Ezan, E., and Junot, C. (2008). Direct introduction of biological samples into a LTQ-Orbitrap hybrid mass spectrometer as a tool for fast metabolome analysis. *Anal. Chem.* **80**, 3291–3303.

Maharjan, R. P. and Ferenci, T. (2003). Global metabolite analysis: the influence of extraction methodology on metabolome profiles of Escherichia coli. *Anal. Biochem.* **313**, 145–154.

Mashego, M. R., Wu, L., Van Dam, J. C., Ras, C., Vinke, J. L., Van Winden, W. A., Van Gulik, W. M., and Heijnen, J. J. (2004). MIRACLE: mass isotopomer ratio analysis of U-13C-labeled extracts. A new method for accurate quantification of changes in concentrations of intracellular metabolites. *Biotechnol. Bioeng.* **85**, 620–628.

Orth, J. D., Conrad, T. M., Na, J., Lerman, J. A., Nam, H., Feist, A. M., and Palsson, B.Ø. (2011). A comprehensive genome-scale reconstruction of Escherichia coli metabolism—2011. *Mol. Syst. Biol.* **7**, 535.

Rabinowitz, J. D. (2007). Cellular metabolomics of *Escherichia coli*. *Expert Rev. Proteomics* **4**, 187–198.

Rabinowitz, J. D. and Kimball, E. (2007). Acidic acetonitrile for cellular metabolome extraction from Escherichia coli. *Anal. Chem.* **79**, 6167–6173.

Schaefer, U., Boos, W., Takors, R., and Weuster-Botz, D. (1999). Automated sampling device for monitoring intracellular metabolite dynamics. *Anal. Biochem.* **270**, 88–96.

Schaub, J., Schiesling, C., Reuss, M., and Dauner, M. (2006). Integrated sampling procedure for metabolome analysis. *Biotechnol. Prog.* **22**, 1434–1442.

Taymaz-Nikerel, H., de Mey, M., Ras, C., ten Pierick, A., Seifar, R. M., van Dam, J. C., Heijnen, J. J., and van Gulik, W. M. (2009). Development and application of a differential method for reliable metabolome analysis in Escherichia coli. *Anal. Biochem.* **386**, 9–19.

van der Werf, M. J., Overkamp, K. M., Muilwijk, B., Coulier, L., and Hankemeier, T. (2007a). Microbial metabolomics: toward a platform with full metabolome coverage. *Anal. Biochem.* **370**, 17–25.

van der Werf, M. J., Takors, R., Smedsgaard, J., Nielsen, J., Ferenci, T., Portais, J. C., Wittmann, C., Hooks, M. A., Tomassini, A., Oldiges, M., Fostel, J., and Sauer, U. (2007b). Standard reporting requirements for biological samples in metabolomics experiments: microbial and in vitro biology experiments. *Metabolomics* **3**, 189–194.

Villas-Bôas, S. G. and Bruheim, P. (2007). Cold glycerol-saline: the promising quenching solution for accurate intracellular metabolite analysis of microbial cells. *Anal. Biochem.* **370**, 87–97.

Visser, D., van Zuylen, G., van Dam, J. C., et al. (2002). Rapid sampling for analysis of in vivo kinetics using the BioScope: a system for continuous-pulse experiments. *Biotechnol. Bioeng.* **79**(6), 674–681.

Volkmer, B. and Heinemann, M. (2011). Condition-dependent cell volume and concentration of Escherichia coli to facilitate data conversion for systems biology modeling. *PLoS One* **6**, e23126.

Weuster-Botz, D. (1997). Sampling tube device for monitoring intracellular metabolite dynamics. *Anal. Biochem.* **246**, 225–233.

Winder, C. L., Dunn, W. B., Schuler, S., et al. (2008). Global metabolic profiling of *Escherichia coli* cultures: an evaluation of methods for quenching and extraction of intracellular metabolites. *Anal. Chem.* **80**, 2939–2948.

Wittmann, C., Krömer, J. O., Kiefer, P., Binz, T., and Heinzle, E. (2004). Impact of the cold shock phenomenon on quantification of intracellular metabolites in bacteria. *Anal. Biochem.* **327**, 135–139.

Wu, L., Mashego, M. R., van Dam, J. C., Proell, A. M., Vinke, J. L., Ras, C., van Winden, W. A., van Gulik, W. M., and Heijnen, J. J. (2005). Quantitative analysis of the microbial metabolome by isotope dilution mass spectrometry using uniformly ^{13}C-labeled cell extracts as internal standards. *Anal. Biochem.* **336**, 164–171.

Yuan, J., Doucette, C. D., Fowler, W. U., et al. (2009). Metabolomics-driven quantitative analysis of ammonia assimilation in E. coli. *Mol. Syst. Biol.* **5**, 302.

CHAPTER

Array-based approaches to bacterial transcriptome analysis

Ulrike Mäder[*,1], Pierre Nicolas[†]

[*]Ernst-Moritz-Arndt-University Greifswald, Interfaculty Institute for Genetics and Functional Genomics, Greifswald, Germany
[†]INRA, UR1077 Mathématique Informatique et Génome (MIG), Domaine de Vilvert, Jouy-en-Josas, France
[1]Corresponding author. e-mail address: ulrike.maeder@uni-greifswald.de

1 INTRODUCTION

Systems level approaches to understand bacterial metabolism and its regulation in response to environmental factors require quantitative, time-resolved data on the abundance of the cellular components, including mRNAs, proteins, and metabolites. In bacteria, changes in transcript levels are a key factor in the adaptive response to changes in environmental conditions. Thus, the analysis of the transcriptome had become a particularly important part of functional genomics and, in the systems biology era, of multi-omics studies aimed at an integrated analysis of the multiple layers of biological information (Zhang et al., 2010). Recent examples include multi-laboratory efforts to analyse adaptations of metabolism to genetic perturbation in yeast (Canelas et al., 2010) and to environmental changes in the Gram-positive model bacterium *Bacillus subtilis* (Buescher et al., 2012).

Measuring amounts of transcripts at a biologically relevant level of precision is, at the time of writing, an easier method for acquiring data on cell physiology than any other systems-level approach. Established techniques for measuring mRNA abundance such as Northern blotting and real-time (quantitative) PCR allow for the analysis of limited numbers of genes (Bustin et al., 2005). However, with the introduction of methods for genome-wide expression analysis, in particular, serial analysis of gene expression (Velculescu et al., 1995) and DNA microarrays (Schena et al., 1995), it became possible to determine the abundance of thousands of mRNA species simultaneously. In the case of DNA microarrays, gene-specific probes are attached to a solid surface by employing different array production technologies. For the analysis, experimental samples are fluorescently labelled and hybridized to the array. After fluorescence detection by means of a laser scanning densitometer, the amount of target hybridized to each oligonucleotide probe can be quantified from the obtained image. Over the past 15 years, microarray technology has been extensively used to compare or quantify genome-wide mRNA levels in both eukaryotic and prokaryotic organisms.

The first purpose of a transcriptome study is often to identify genes that are differentially expressed between two or more conditions, and statistical testing for differential expression of genes is now a relatively streamlined procedure. However, making biological sense of lists of differentially expressed genes is often not trivial and many commonly applied analytical approaches extend beyond testing for the differential expression of single genes. This includes, for example, the analysis of the expression patterns of predefined gene sets (groups of functionally related genes; reviewed in Ackermann and Strimmer, 2009) or the application of clustering methods whose purpose is to delineate sets of genes that share similar expression patterns across the experimental conditions and thereby facilitating the tentative assignment of functions to previously uncharacterized genes (reviewed by Quackenbush, 2001). Furthermore, the availability of data sets collected in different genetic backgrounds and under a number of experimental conditions initiated the development of methods to infer transcriptional regulatory networks, most often combining cluster analysis of gene expression data with transcription factor binding motif searches (reviewed by Herrgård et al., 2004).

Recent technical advances in high-density array formats and DNA sequencing technologies (i.e. next generation sequencing, NGS) associated with the growing number of sequenced bacterial genomes have opened the way to the characterization of entire transcriptomes by genomic tiling arrays and whole-transcriptome sequencing (RNA-seq) methods. The rapid development of these powerful tools for unbiased transcriptome analysis is reflected in several recent reviews on prokaryotic transcriptomics (e.g. Sorek and Cossart, 2010; Mäder et al., 2011; Filiatrault, 2011). Genomic tiling arrays, which are high-density oligonucleotide arrays covering both strands of a genome by overlapping probes, facilitate the mapping and quantification of all transcribed regions with a resolution determined by the distance between two adjacent probes which is typically 10–20 nucleotides in the case of microorganisms. Whole-transcriptome studies based on tiling arrays and/or RNA-seq have revealed an unexpectedly high level of complexity within the bacterial transcriptome (Sorek and Cossart, 2010), even when only a relatively small number of experimental conditions are analysed. A recent tiling array study of *B. subtilis* grown under a wide range of nutritional and environmental conditions has established a comprehensive repertoire of transcription units and identified, for example, more than 1500 new RNA features including 512 potentially new genes (Nicolas et al., 2012). For the first time, this study evaluated the contribution of a bacterium's RNA polymerase sigma factors to global transcriptional regulation, an analysis made possible by the availability of promoter-level data with high signal-to-noise ratio.

Systems biology approaches require the comprehensive identification and precise quantification of all transcripts of a given sample. In line with this, genomic tiling arrays can be used to detect and annotate novel transcripts including 5'- and 3'-untranslated regions, protein-encoding mRNAs as well as regulatory and catalytic RNAs, thereby expanding the existing genome annotation. Furthermore, they facilitate the mapping of transcription units, the detection of alternative promoters used under different experimental conditions, and the quantitation of transcript abundance. Lastly, tiling arrays can be used for the genome-wide mapping of protein–DNA interactions by chromatin immunoprecipitation (ChIP)-on-chip analysis (reviewed by Buck and Lieb, 2004). On the other hand, classical microarrays (so-called gene expression arrays), which are based on an existing genome annotation and contain relatively few probes for each gene, are well suited to assess the expression levels of annotated transcripts under many different conditions. They are less expensive and the data easier to analyse than genomic tiling arrays.

In this chapter, we give a brief overview of microarray platform and design options. Further, we describe well-established workflows for bacterial transcriptome analysis including the preparation of high-quality RNA samples, cDNA labelling and array processing. Finally, Section 4 describes and discusses succinctly the steps that lead to the biological interpretation of the expression profiles: from the normalization and probe aggregation methods to the differential expression assessment and gene regulatory networks analyses.

2 PRIOR CONSIDERATIONS
2.1 Technical requirements for microarray experiments
2.1.1 Microarray platforms

With respect to the array fabrication methods, the microarray field was initially dominated by two major technologies, namely, spotted array systems and Affymetrix GeneChip technology (photolithographic *in situ* synthesis of short oligonucleotides). Spotted arrays are produced by depositing pre-synthesized oligonucleotides (also cDNA or PCR products) on a glass slide. During the past decade, important technological developments in the field of *in situ* synthesized arrays, introduced, in particular, by NimbleGen (Maskless Array Synthesizer technology), Agilent Technologies (SurePrint inkjet technology), and Illumina (BeadArray technology), have greatly improved the sensitivity (use of longer oligonucleotides), design flexibility (maskless probe synthesis), reproducibility, and cost–effectiveness of microarray experiments. Importantly, higher probe densities of *in situ* synthesized arrays have enabled tiling designs covering both strands of a given genome by overlapping probes. A single slide is sufficient to tile a bacterial genome at high resolution, that is, with a tiling step of less than 10 nucleotides (nt).

Because of the reproducibility of the manufacturing process, which reduces the variability between individual slides, *in situ* synthesized arrays, unlike spotted arrays, are not restricted to two-colour competitive hybridizations where sample and control (or common reference) cDNAs are labelled with different fluorescent dyes (commonly, Cy5 and Cy3) and hybridized to the same array. The relative abundance of a transcript in the two samples is determined by the Cy5/Cy3 signal ratio of the corresponding spot. The two-colour approach can reliably measure the difference between two samples, as variations in spot size or amount of probe on the array do not affect the signal ratio. However, as the analysis is always based on direct comparisons between the co-hybridized samples, the experimental design can be quite complicated and inflexible for large sample sets, which is especially true for interwoven loop designs (Churchill, 2002; Khanin and Wit, 2005). In the case of *in situ* synthesized arrays, the low variability between individual slides is associated with an improved reproducibility of the data at the signal level thus allowing for one-colour hybridizations (Patterson *et al.*, 2006). Unlike the Affymetrix system, which is solely based on single-channel detection, the NimbleGen and Agilent platforms allow for both one- and two-colour approaches. Two-colour, common reference designs can be an advantage with respect to controlling the variation associated with the labelling and hybridization procedures but, on the other hand, do not allow for direct comparisons of data from different studies. The common reference approach using a pool of all samples included in a study will be discussed in Section 2.2.

Because of the limitations of spotted arrays, such as higher slide-to-slide variation, lower probe density and less flexibility with regard to probe content, only *in situ* synthesized arrays will be considered in this chapter. However, this does not preclude the use of spotted arrays for generating reliable data if the technical performance of array production and processing is sufficiently well controlled and adequate care is taken in experimental design and data analysis. Two recent examples of such studies have been published by Hahne *et al.* (2010) and Blom *et al.* (2011).

2.1.2 Array designs and array processing requirements

The largest collection of bacterial expression microarray designs (so-called catalogue arrays) is offered by Roche NimbleGen, where designs for over 200 bacterial species are currently available, but will most probably cease to be produced by the end of 2012. For *Escherichia coli*, a tiling array design is also offered by Roche NimbleGen containing 385,000 probes and having a median probe spacing of 24 nt. In addition, expression microarrays for many bacterial species are available from MYcroarray, a company employing maskless photolithography, or Genotypic that uses Agilent's SurePrint *in situ* technology. Agilent's collection of model organism gene expression microarrays includes an *E. coli* array containing 15,000 probes, and Affymetrix's previous generation GeneChip arrays can be obtained as pre-designed bacterial arrays for *E. coli*, *B. subtilis*, and *Staphylococcus aureus*. The approach applied by most researchers, particularly with respect to genomic tiling arrays, is to perform the probe design for the studied organism and then have the arrays produced by a commercial manufacturer (so-called custom-designed arrays). In contrast to spotted arrays, *in situ* synthesized, high-density arrays depend on commercial manufacturing because of the specific production processes employed.

There are many aspects of the probe design that have to be taken into consideration in order to generate high-quality experimental data. Importantly, the thermodynamic properties of all probes on the array should be as similar as possible. Probe quality parameters include melting temperature (adjustment of probe length for obtaining a near-isothermal design, i.e. homogeneous T_m which correlates to probe affinity), potential secondary structures, cross-hybridization and probe placement; probe design software programs such as OligoWiz (Nielsen *et al.*, 2003) take these parameters into account. The number of probes is obviously determined by the application (expression microarray vs. genomic tiling array), the size of the genome and the array formats provided by the manufacturer. For example, a *B. subtilis* genomic tiling array has been designed using OligoWiz 2.0 (Wernersson and Nielsen, 2005) and consists of isothermal probes (45–65 nt in length) starting every 22 nt on both strands. The design contains a total of 385,000 probes matching the NimbleGen 385K array format (Rasmussen *et al.*, 2009).

For the design of tiling array probes, the web-based OligoWiz program, which was initially developed for expression microarray probe selection, has been successfully employed (Rasmussen *et al.*, 2009; Thomassen *et al.*, 2009). Recently, chipD has been created for the design of high-density tiling array probes; the program can account for specific applications, protocols or array formats (Dufour *et al.*, 2010) and has been used to design bacterial tiling arrays (Peters *et al.*, 2009). Another web tool for probe selection for prokaryotic tiling arrays has been developed by Høvik and Chen (2010). A number of software programs available for the design of oligonucleotide probes for expression and tiling arrays have been compared and evaluated by Lemoine *et al.* (2009).

In addition to array design, the specific equipment required for the hybridization, washing and scanning of the array slides has to be considered. Microarray processing instruments are produced by several companies, and their compatibility with the array fabrication technology, array format and processing protocols needs to be confirmed. If the purchase of costly instruments or the training of personnel is not economically justified (e.g. for a small number of experiments), certified service providers for each of the common platforms can provide suitable, high-quality alternatives to in-house sample processing. The Agilent Technologies and Roche NimbleGen platforms are most relevant with respect to bacterial expression microarrays as well as genomic tiling arrays, and these will be specifically addressed in this chapter. In addition to high-density arrays, both companies offer a complete workflow for DNA microarray processing, including a hybridization oven or hybridization station, an optional slide washing station and a high-resolution microarray scanner with a pixel resolution of 2 μm. These instruments can also handle slides from other platforms, whereas the Affymetrix and Illumina systems are so-called closed platforms, intended for the exclusive use with the corresponding arrays. It is also possible to implement semi-automated solutions as, for example, the HS Pro hybridization stations from Tecan for hybridization and washing. Specific protocols for one- and two-colour gene expression analysis involving these systems are provided by Agilent.

2.1.3 Software requirements

The most popular software for microarray data analysis in the computational biology community is R (http://www.r-project.org/) that provides a free environment for statistical computing and graphics with precompiled versions available for Windows,

MacOS X and Linux. Alongside its base system that provides very generic functions, R is being continuously enriched by packages that can be developed by any researcher. These packages are available on the CRAN (The Comprehensive R Archive Network) or through the more specialized Bioconductor project that coordinates a large number of open source R package contributions in the field of computational biology (Gentleman et al., 2004). A number of approaches, which are often computationally intensive and therefore not well suited for interactive use with R, are distributed as standalone applications. An alternative to R is provided by commercial software such as GeneSpring (distributed by Agilent Technologies) offering environments for microarray data analysis with more intuitive interfaces that do not require programming skills and allow guided but yet somewhat flexible analysis workflows.

2.2 Experimental design

When designing a microarray study, the experimental details including data analysis procedures and, if applicable, integration with data from other ('omics) approaches need to be dealt with carefully in order to ensure that the generated data will address the relevant biological questions. More specifically, given the amount of work and resources needed to perform the experiments, several issues crucial for obtaining most informative results should be considered from the outset. These include the data type (expression microarray vs. genomic tiling array), the experiment type (one-colour vs. two-colour hybridizations), the number of biological replicates and the potential batch effects (Leek et al., 2010). Certainly, quantifying (changes in) transcript abundance under different experimental conditions and/or time-points is usually the main objective of a transcriptome study. An additional substantial aspect of unbiased transcriptomics approaches is to detect and annotate transcribed regions that lie outside the current genome annotation. Mainly because of the latter aspect, tiling arrays often outperform classical microarrays in systems biology studies. However, in view of constraints such as costs or availability of probe design and data analysis capabilities, expression microarrays remain a useful tool for the analysis of annotated transcripts. For a series of studies involving a single bacterial species, a suitable approach could be first to apply genomic tiling arrays to identify novel transcripts systematically under a wide range of experimental conditions, and then to use the resulting improved annotation for the design of a comprehensive but cheaper gene expression microarray.

As mentioned before, two-colour hybridizations are generally used for spotted arrays because of variations in spot size and/or amount of probes. However, with *in situ* synthesized arrays, it can still be advantageous to perform competitive hybridizations, in particular, using common reference designs that do not require special analysis procedures and can be extended as long as the reference RNA is available. The preferred choice for the common reference RNA is a pool of all samples included in the study that provides an average expression profile. This requires the gene ratios for the individual samples to reflect the expression levels relative to the average across all samples. In two-colour experiments, the hybridization signals of paired samples are compared thereby controlling for potential variation introduced during sample labelling and array processing. Thus, the advantage of this approach over one-colour experiments is its robustness in the face of day-to-day variations. It is, however, crucial that the reference sample is available in sufficient amounts and consistent quality throughout the course of the entire study.

The availability of data on biological replicates, consisting of independent cultivations of the same genetic background in the same growth conditions, is important for most applications in microbial transcriptomics, and crucial if the goal is to identify differential expression between different culture conditions and genetic backgrounds. More replicates provide higher statistical power but directly increase the cost of the experiment, thereby limiting the number of conditions/strains that can be explored. More precisely, as discussed in Bolstad *et al.* (2004), the statistical power to detect differential expression depends on the level of variance between replicates and the proportion and fold-change amplitude of the differentially expressed genes. Therefore, the choice of the number of replicates may seem important but the information needed for making a really wise choice is, in general, not available beforehand. In practice, if the experimental set-up achieves low variance between replicates, a design involving three biological replicates will already offer sufficient statistical power to detect biologically relevant changes (i.e. either with important fold changes or involving a large number of genes) and can therefore be considered as a reasonable starting point. Of note, biological replicates account for both the biological and technical sources of variation. In contrast, technical replicates would be helpful only to delineate the respective share of the biological and technical variances and are usually not included in the experimental design.

Another important issue that needs to be considered is between-sample normalization. Global methods such as quantile normalization or median centring are employed in the majority of transcriptome studies and will be elaborated in Section 4.1. However, one should be aware that such approaches rely on the assumption that the distribution of expression levels is, at least to some extent, preserved under the conditions to be compared. This will be true if the expression level of most genes is unchanged and if the pattern of up- and down-regulated genes is roughly symmetric. In the case of global changes in gene expression, these normalization methods (based on endogenous transcript levels) will not accurately assess differential expression. Indeed, if global changes between samples are expected, an alternative is external normalization based on a strategy of "spike-in" controls added in equal amounts to each total RNA sample before cDNA synthesis (van de Peppel *et al.*, 2003). This needs to be considered during the design or selection of the array to ensure that it contains the appropriate control probes (Thomassen *et al.*, 2009). For the Agilent platform, spike-in controls are available as mixtures of 10 different *in vitro* synthesized transcripts that hybridize to complementary control probes on the arrays. By adding spike-in controls to each total RNA sample, the average difference of the control signal intensities between different arrays can be used to scale the intensities of each individual array (Chandriani and Ganem, 2007). However, further calculation steps are required if not only the mRNA amount but also the amount of total RNA per cell changes between the biological conditions compared.

3 PERFORMING THE EXPERIMENTS
3.1 RNA preparation and quality assessment

The preparation of high-quality RNA is an extremely important prerequisite for generating reliable transcriptome data, even more so for bacteria, where mRNA half-lives are short (mostly in the range of a few minutes). The different methods employed usually follow one of three general techniques: (i) organic extraction using acid phenol or commercially available solutions of guanidine isothiocyanate and phenol (TRIzol, TriFast or TRI Reagent) based on the single-step RNA isolation method developed by Chomczynski and Sacchi (1987), (ii) adsorption of RNA onto a spin column membrane (silica matrix or glass fibre filter) of commercially available kits (NB. For isolating total RNA including small RNAs < 200 nt spin columns need to be combined with organic extraction, except for Norgen's Total RNA Purification Kit.) and (iii) magnetic particle-based (automated) methods. In each case, the first issue to be considered is maintaining RNA integrity prior to RNA isolation, that is, during and after harvesting the bacteria. An appropriate procedure involves centrifugation at low temperature, freezing the cell pellets in liquid nitrogen and storing them at $-80\,°C$ until cell disruption. To prevent changes in the transcriptome due to stress responses induced during centrifugation, sodium azide is added to the harvesting buffer (see Winter et al., 2011).

At the beginning of each RNA preparation procedure, bacterial cell lysates are prepared, usually by enzymatic lysis or mechanical disruption. Efficient cell disruption while maintaining RNA integrity is particularly an issue for Gram-positive bacteria. The method recommended involves mechanical cell disruption followed by organic extraction with acid phenol (Eymann et al., 2002). This method, originally developed for preparation of high-quality RNA from yeast (Hauser et al., 1998), has, over the past 10 years, been extensively used in studies involving B. subtilis, Bacillus anthracis and S. aureus. A detailed protocol is provided in Nicolas et al. (2012). Cells frozen in liquid nitrogen are rapidly disrupted by means of a grinding ball in a Mikro-Dismembrator instrument (Sartorius) followed by completion of lysis with guanidine thiocyanate as denaturing agent. This procedure effectively protects the RNA from degradation, yielding RNA preparations with a high degree of integrity as indicated by rRNA precursor bands visible above the 23S rRNA band in gel electrophoresis (Figure 6.1). In general, it is important that the RNA preparation method results in the complete size range of RNA (not selectively excluding small or large transcripts) and that the same method is used for all samples to be compared because different methods are likely to have an impact on the transcriptome.

Before the RNA samples can be used for transcriptome analysis, residual DNA has to be removed by DNase treatment using, for example, the RNase-Free DNase

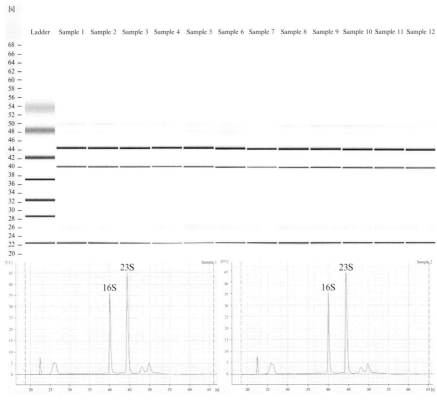

FIGURE 6.1

Gel-like image and example electropherograms of *B. subtilis* total RNA samples analyzed using an RNA 6000 Nano LabChip Kit (Agilent Technologies) on the Agilent 2100 Bioanalyzer. The high degree of integrity of the RNA preparations is indicated by the rRNA precursor bands visible above the 23S rRNA band. FU, fluorescence units. (For colour version of this figure, the reader is referred to the online version of this chapter.)

Set (Qiagen). Subsequently, the samples are purified and the spin column-based RNA Clean-Up and Concentration Kit from Norgen Biotek has been shown to retain the small RNA species. RNA concentration and purity are then determined spectrophotometrically, if possible by means of a NanoDrop instrument (Thermo Scientific). RNA purity is a critical factor because contaminants can strongly impair the efficiency of cDNA synthesis and labelling; therefore, the absorbance ratio of A_{260nm}/A_{280nm} should be >2 (low values indicate protein or phenol contamination) and the absorbance ratio of A_{260nm}/A_{230nm} should be >1.8 (low values can indicate various compounds such as phenol, guanidine thiocyanate or polysaccharides). Finally, the integrity of the RNA preparations needs to be validated using the Agilent Bioanalyzer (Agilent Technologies) or similar instrument, used according to the manufacturer's instructions (Figure 6.1). During the past decade, the Agilent Bioanalyzer has become the standard tool for assessing RNA quality, particularly for RNA from eukaryotic cells where, in addition to the electropherograms, RIN (RNA Integrity Number) values can be used for an objective and standardized evaluation of RNA quality.

3.2 Synthesis and labelling of cDNA

In typical array-based bacterial transcriptome studies, total RNA is converted into first-strand cDNA using random primers and reverse transcriptase. Depending on the scanning technology employed, cDNA is labelled either with a fluorescent dye or with biotin (Affymetrix GeneChip platform). For fluorescence labelling, two basic principles can be applied either direct labelling (reverse transcription with concomitant incorporation of a cyanine-dye-coupled nucleotide) or indirect labelling *via* aminoallyl nucleotides (incorporation of an aminoallyl-modified nucleotide and subsequent chemical coupling to a cyanine dye). Possible disadvantages of direct labelling procedures such as lower cDNA yields and less efficient incorporation of labelled nucleotides have been discussed (Richter *et al.*, 2002) but cannot be confirmed on the basis of our current experience. Before being used for hybridization,

---

Protocol 1 GENERATION AND LABELLING OF STRAND-SPECIFIC CDNA FROM PROKARYOTIC TOTAL RNA

Specific Reagents: FairPlay III Microarray Labeling Kit (Stratagene) from Agilent Technologies, Actinomycin D (1 set = 20 × 200 µg) from Calbiochem, CyDye Cy3 Mono-Reactive Dye 5-Pack from GE Healthcare

Step 1. First-Strand cDNA Synthesis
1. Prepare Actinomycin D by dissolving 200 µg (entire vial) in 400 µl RNase/DNase-free water. Make aliquots of 20 µl and store at −20 °C. These aliquots are stable for at most a month at −20 °C.
2. Assemble the following components in 0.2-ml thin-walled PCR tubes: 9 µl sample (10 µg RNA) and 1 µl Random Primer (9 mers, 500 ng/µl).
3. Heat-denature sample in a thermocycler at 70 °C for 10 min. Quick-chill in an ice-water bath for 5 min.
4. Prepare the following first-strand Master Mix according to the number of reactions (volumes per sample):

10× AffinityScript RT buffer	2 µl
20× dNTP mix	1 µl
0.1 M DTT	1.5 µl
RNase Block (40 U/µl)	0.5 µl
Actinomycin D (500 µg/ml)	1.6 µl
RNase/DNase-free water	0.4 µl
Total	7 µl

5. Add 7 μl of the first-strand Master Mix to each denatured sample from step 1.3. Keep on ice.
6. Mix well by pipetting 10 times.
7. Quick-spin to force contents to bottom of tube.
8. Add 3 μl of AffinityScript HC Reverse Transcriptase to each reaction (final volume of 20 μl).
9. Incubate at room temperature for 1 h.
10. Incubate in a thermocycler at 42 °C for 1 h.
11. Add 10 μl of 1 M NaOH to each reaction and incubate at 70 °C for 10 min to hydrolyze RNA.
12. Cool to room temperature slowly; do not cool on ice.
13. Centrifuge the tube briefly to collect contents.
14. Add 10 μl of 1 M HCl to neutralize the solution.

Step 2. cDNA Purification
1. Quick-spin and add 4 μl of ice-cold 7.5 M ammonium acetate and 1 ml 20 mg/ml glycogen to each reaction from step 1.14.
2. Add 100 μl of ice-cold 100% ethanol. Mix well by pipetting and transfer to a 1.5-ml tube.
3. Incubate at −20 °C overnight. The reaction can be stored at this point for several days or up to 2 months.
4. Centrifuge the reaction from step 2.3 at 13,000–14,000 × g for 15 min at 4 C. Carefully decant supernatant.
5. Wash with 0.5 ml ice-cold 70% ethanol and centrifuge at 13,000–14,000 × g for 15 min at 4 °C. Carefully decant supernatant and allow to air dry or dry in vacuum dryer. Do NOT overdry.

Step 3. NHS-Ester Dye Coupling Reaction
1. Resuspend the cDNA pellet in 5 μl of 2 × coupling buffer. If a precipitate can be seen in the 2 × coupling buffer, incubate the buffer at room temperature or 37 °C to re-solubilize the precipitate before use.
2. The first time a tube of dye is used, resuspend in 45 μl of the high-purity DMSO provided in the kit. Vortex gently to ensure the pellet is completely solubilized. Prepare single use aliquots of the unused dye, which can be stored at −20 °C in the dark for several months.
3. Add 5 μl of Cy3 dye to the cDNA. If the dye was stored at −20 °C prior to use, allow the dye to reach room temperature before opening the container.
4. Mix by gently pipetting up and down.
5. Incubate for 30 min at room temperature in the dark.

Step 4. Dye-Coupled cDNA Purification
1. Add 90 μl of RNase/DNase-free water to the 10 μl labelled cDNA from step 3.5.
2. Combine 100 μl of DNA-binding solution and 100 μl of 70% (v/v) ethanol. Mix well by vortexing. Make sure that the two solutions are well mixed prior to use.
3. Add the 200 μl of DNA-binding solution/ethanol mixture to the labelled cDNA from step 4.1 and mix by vortexing.
4. Using a pipette, transfer the mixture directly on the filter of a microspin cup that is seated in a 2-ml receptacle tube. (Exercise caution to avoid damaging the fibre matrix with the pipette tip.) Snap the cap of the 2-ml receptacle tube onto the top of the microspin cup. Note: To ensure proper sample flow, use the receptacle tube that is provided with the microspin cups.
5. Centrifuge the tube in a microcentrifuge at maximum speed for 30 s. Note: The labelled cDNA is retained in the fibre matrix of the microspin cup.

Continued

> **Protocol 1 GENERATION AND LABELLING OF STRAND-SPECIFIC cDNA FROM PROKARYOTIC TOTAL RNA—Cont'd**
>
> 6. Open the cap of the 2-ml receptacle tube, remove and retain the microspin cup, and discard the DNA-binding solution containing the uncoupled dye.
> 7. Combine 100 μl of the DNA-binding solution and 100 μl of 70% (v/v) ethanol. Mix well by vortexing. Make sure that the two solutions are well mixed prior to use.
> 8. Add the 200 μl of DNA-binding solution/ethanol mixture to the microspin cup. Snap the cap of the receptacle tube onto the top of the microspin cup.
> 9. Centrifuge the tube in a microcentrifuge at maximum speed for 30 s.
> 10. Open the cap of the 2-ml receptacle tube, remove and retain the microspin cup, and discard the DNA-binding solution/ethanol mixture.
> 11. Add 750 μl of 75% ethanol to the microspin cup. Snap the cap of the receptacle tube onto the top of the microspin cup.
> 12. Centrifuge the tube in a microcentrifuge at maximum speed for 30 s.
> 13. Open the cap of the 2-ml receptacle tube, remove and retain the microspin cup, and discard the wash buffer.
> 14. Repeat steps 11–13.
> 15. Place the microspin cup back in the 2-ml receptacle tube and snap the cap onto the microspin cup.
> 16. Centrifuge the tube in a microcentrifuge at maximum speed for 1 min.
> 17. Transfer the microspin cup to a fresh 1.5-ml microcentrifuge tube and discard the 2-ml receptacle tube.
> 18. Add 50 μl of 10 mM Tris base, pH 8.5 directly onto the top of the fibre matrix at the bottom of the microspin cup.
> 19. Incubate the tube at room temperature, in the dark, for 5 min.
> 20. Snap the cap of the 1.5-ml microcentrifuge tube onto the microspin cup and centrifuge the tube in a microcentrifuge at maximum speed for 30 s.
> 21. Open the lid of the microcentrifuge tube and recover the flow through containing the purified labelled cDNA.
> 22. Elute additional labelled cDNA by pipetting the flow through back onto the fibre matrix of the same microspin cup.
> 23. Re-seat the spin cup on the same 2-ml receptacle tube that contained the liquid from the first-pass elution.
> 24. Incubate the tube at room temperature for 5 min.
> 25. Snap the cap of the 1.5-ml microcentrifuge tube onto the microspin cup and centrifuge the tube in a microcentrifuge at maximum speed for 30 s.
> 26. Open the lid of the microcentrifuge tube and recover the flow through containing the purified labelled cDNA.
> 27. Harvest one final elution from the microspin cup by repeating steps 22–26.
> 28. Open the lid of the microcentrifuge tube and recover the flow through containing the purified labelled cDNA.
>
> Step 5. Spectrophotometric Quantitation and Quality Control of cDNA
> 1. Determine cDNA and Cy3 dye concentrations by using a small-volume spectrophotometer (such as a NanoDrop instrument).
> 2. Quantitate each cDNA sample according to the following formula:
>   cDNA concentration (ng/μl) = $A_{260} \times 33 \times$ Dilution Factor
> 3. Verify that all samples meet the following requirements:
>   - Concentration ≥ 20 ng/μl
>   - $A_{260}/A_{280} \geq 1.8$
>   - $A_{260}/A_{230} \geq 1.8$

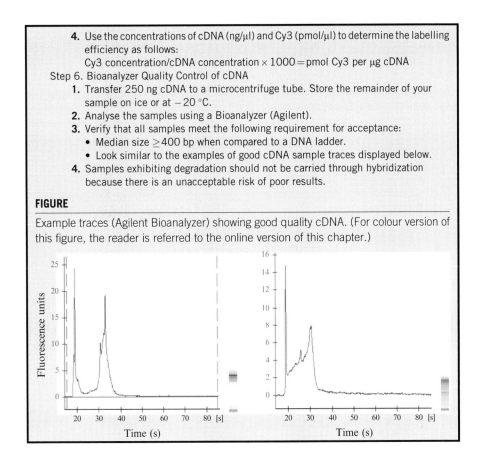

4. Use the concentrations of cDNA (ng/μl) and Cy3 (pmol/μl) to determine the labelling efficiency as follows:
   Cy3 concentration/cDNA concentration × 1000 = pmol Cy3 per μg cDNA

Step 6. Bioanalyzer Quality Control of cDNA

1. Transfer 250 ng cDNA to a microcentrifuge tube. Store the remainder of your sample on ice or at −20 °C.
2. Analyse the samples using a Bioanalyzer (Agilent).
3. Verify that all samples meet the following requirement for acceptance:
   - Median size ≥400 bp when compared to a DNA ladder.
   - Look similar to the examples of good cDNA sample traces displayed below.
4. Samples exhibiting degradation should not be carried through hybridization because there is an unacceptable risk of poor results.

FIGURE

Example traces (Agilent Bioanalyzer) showing good quality cDNA. (For colour version of this figure, the reader is referred to the online version of this chapter.)

the dye-labelled cDNA is purified and quality controlled to monitor the performance of the labelling procedure. Purification of cDNA is usually performed with spin-column based kits, for example, the CyScribe GFX Purification Kit (GE Healthcare). In the next step, the cDNA yield and labelling efficiency are assessed using a NanoDrop spectrophotometer (Thermo Scientific) following the manufacturer's instructions. As a further quality parameter, the size distribution of the cDNA samples can be analysed (see protocol below). If the quality measures do not match the thresholds specified in the employed protocol or established during protocol development, the affected cDNA samples should not be subjected to hybridization.

It is essential to note that reverse transcriptase generates spurious second-strand cDNA due to its DNA-dependent polymerase activity (Ruprecht et al., 1973). As second-strand synthesis leads to antisense artefacts as well biases in the quantification of expression levels, it is important to inhibit the DNA-dependent polymerase activity of the reverse transcriptase with actinomycin D, as first shown by Perocchi et al. (2007) for the example of yeast tiling array experiments. As an alternative approach to circumventing potential artefacts associated with reverse transcription, direct RNA labelling methods have occasionally been employed (Georg et al., 2009; Yu et al., 2011). A protocol for the generation of strand-specific cDNA for tiling array hybridization using the FairPlay III Microarray Labeling Kit (Stratagene) was developed by Roche NimbleGen within the European BaSysBio project (Rasmussen et al., 2009). This method (Protocol 1) involves incorporation of aminoallyl-dUTP during reverse transcription in the presence of actinomycin D and subsequent labelling of the cDNA with Cy3. It has been successfully applied in a number of recent tiling arrays studies involving *B. subtilis* and *S. aureus* (Rasmussen et al., 2009; Falord et al., 2011; Buescher et al., 2012; Nicolas et al., 2012).

3.3 Array hybridization, scanning and data extraction

In the next steps, the labelled cDNA samples are hybridized to the arrays (usually overnight), and the slides are then washed and scanned according to the platform specific protocols and using the respective instruments as discussed in Section 2. For two-colour competitive hybridizations sample and reference, cDNAs (labelled with Cy5 and Cy3, respectively) are mixed in equal amounts and hybridized to the same array. As washing artefacts can cause problems by introducing background variability, the washing procedure is crucial for the quality of the resulting data. When scanning the arrays, attention needs to be paid to selecting an appropriate photo-multiplier tube (PMT) settings so that only very few or no probes will be saturated in order to avoid quantitation errors. However, those lower sensitivity settings have the obvious disadvantage of limiting the ability to reliably detect weak signals. Importantly, a wider dynamic range is obtained by the extended dynamic range (XDR) scanning implemented in the Agilent scan control software, thus covering a larger range of transcript abundance. In XDR mode, the scanner performs a dual scan (at 100% and 10% PMT, respectively) and the two images are automatically combined when imported into the Feature Extraction software (Agilent Technologies; see next paragraph).

Finally, the data are extracted from the scanned image by using appropriate image analysis software generally supplied with the microarray scanner, but platform-independent solutions are available as well. Data extraction basically involves grid alignment, spot finding, rejection of outlier pixels and some kind of background correction. Additionally, in the case of two-channel arrays, a normalization step is required, typically involving Loess normalization which corrects for intensity-dependent dye bias. As a result of the data extraction process, probe intensities (and ratios) as well as statistical confidence measures associated with each probe intensity (and ratio) are calculated. When working with the Agilent microarray platform, image analysis is usually performed using Feature Extraction software which provides a series of spot quality measures and the special feature of summary values for measurement uncertainty for one-colour arrays (i.e. the processed signal errors) and two-colour arrays (i.e. the log-ratio errors) based on the selected error model.

Having extracted relative expression values in this manner for the individual probes, the values of multiple probes that query the same gene need to be aggregated in order to produce a single gene level value (see Section 4.1). For Agilent expression microarrays, the authors preferentially apply error-weighted averaging to calculate a single expression value for each gene (Weng et al., 2006), using Rosetta Resolver from Rosetta Biosoftware or similar and based on the error measures given by the Feature Extraction software.

4 DATA ANALYSIS

4.1 Data preprocessing and quality checks

The term preprocessing usually refers to the three tasks of background correction, normalization and summarization which are combined in this order in the popular Robust Multichip Average (RMA) approach (Irizarry et al., 2003) that was proposed for one-colour Affymetrix microarrays.

Two-colour arrays usually require within-array normalization to account for dye-bias before computing the log-ratio of the red (R, Cy5) and green (G, Cy3) signals. This is typically performed via Loess regression on the $M = \log2(R) - \log2(G)$ versus $A = 0.5*(\log2(R) + \log2(G))$ plot (Yang et al., 2002). Often, log-ratios obtained after dye-bias correction from two-colour arrays can directly be used in downstream analyses such as differential expression testing and clustering (Bolstad et al., 2004; Grant et al., 2007). However, when the range of the log-ratios varies importantly across replicates, between-array normalization, for instance, scale normalization that makes the variance similar, can be useful (Bolstad et al., 2004). Spotted two-colour arrays should not be used for single-channel analysis due to variations between arrays caused by the manufacturing process that can be controlled for only by comparison of the two signals on the same array. Red and green signals for other two-colour arrays (see Section 2.1.1) can in some cases be analyzed separately but one should be aware that they are not truly independent due to competition between the two samples that become prominent when the signal saturates.

A comparison of workflows commonly employed for one-colour Affymetrix GeneChip arrays (Irizarry et al., 2006) demonstrated the impact of background correction, which aims at subtracting the signal arising from optical and hybridization noise. Background correction wish to decrease bias (i.e. increase accuracy) but all the proposed procedures also tend to increase variance (i.e. decrease precision). The approach can be global in the sense that it assumes a same distribution of background noise for all the probes (e.g. RMA) or local and then try to make probe specific corrections by taking into account the probe characteristics, in terms of nucleotide composition (e.g. GCRMA, Wu et al., 2004), behaviour in control data sets (Huber et al., 2006) or measures reflecting the local background image noise estimated by the image analysis software (Huber et al., 2002). When using array technologies with long probes, typically ~55 nt as compared to ~25 nt for Affymetrix GeneChips, that are more often adjusted to be near-isothermal, the background hybridization noise compared to targeted hybridization signal tends to be lower and background correction is sometimes neglected (Grant et al., 2007). For instance, the standard signal correction method for Agilent expression arrays is "spatial detrending" without background subtraction which consists of two steps: calculating a surface fit that captures the spatial trends of the expression signal (the "foreground") and subtracting the surface fit from the data in order to correct for unwanted spatial heterogeneity. Details of the spatial detrending algorithm are available in the Agilent Feature Extraction v10.7 Reference Guide.

Between-array normalization is often a difficult but important step in one-colour array data preprocessing (Bolstad et al., 2003). The goal here is to filter out the obvious differences in the distribution of the intensity levels from one array to another. These arise from variations in the mRNA extraction, labelling, hybridization (including labelling efficiency and amount of cDNA deposited), washing and also image acquisition (sensitive to the scanner settings). For this purpose, we generally apply some transformation to the log2-scale intensity levels to make them comparable between arrays. The two most popular transformations are the adjustment by a scaling factor (for instance median subtraction on the log2 transformed data) that is a relatively mild transformation and the quantile normalization that consists of matching not only the medians (i.e. the 50%-quantile) but also the whole distributions (i.e. all the quantiles) between samples. These two transformations are global in the sense that the transformation depends on the whole distribution and will therefore tend to smooth out the global trends in the distribution of expression levels. However, this should not be interpreted as removing most of the differences between conditions. Indeed all genes can be differentially expressed, even when the distribution of expression levels is identical. Nevertheless, when global trends are expected, for instance, when experimentally depleting the amount of an RNase and therefore increasing the half-lives of a large number of mRNAs, we would expect most direct effects to be positive and an aggressive global normalization such as quantile normalization may be inappropriate (see Durand et al., 2012). In this case, it can be useful to select a subset (for instance, 10%) of genes that exhibit the smallest variations and learn the transformations making the distribution of expression levels of this subset of genes match across hybridizations. Such procedures are often termed "invariant set" normalizations and can provide satisfying results. However, if one anticipates that global normalization will be inappropriate, the best solution is provided by experimental spike-in of the sample before labelling, as pointed out in Section 2.2. The spike-in signals of the corresponding probes on the array then serve to adjust the transformation applied to the expression levels of each sample. It should be noted, however, that spike-in normalization cannot account for differences in the quality of mRNA and precision of the amount of added spike-in material may be limiting. A final remark about between-array normalization is that it should always put hybridizations from the same and from different conditions on an equal footing: separate normalization for the different conditions would artificially shrink the differences between replicates and amplify the differences between conditions, making the data misleading.

Summarization to probe set level aggregates the intensities of multiple probes that query the same transcript into a single number. This can be achieved by simply taking the median or the average of the log2-values. However, other approaches try to combine information from multiple samples to correct for systematic trends of the different probes that may reflect probe affinity (RMA) or to achieve greater robustness by giving non-equal weights to the different probes (error-weighted mean in the Rosetta Resolver software, Weng *et al.*, 2006; Tukey's biweight robust mean, Hubbell *et al.*, 2002). In the context of long isothermal probes, the variations between probes are more limited and the choice of the aggregation method and its position in the workflow have correspondingly less impact. While aggregation is generally performed after normalization, for tiling array data with long near-isothermal probes (Nicolas *et al.*, 2012), the authors preferred to perform normalization as the last step of the preprocessing workflow. Indeed, in their set-up, the choice of the aggregation method was less critical than the choice of the normalization method and, depending on the question, it might be preferable to work with data that have been normalized using different approaches (see expression vs. differential expression). Being able to redo the normalization without having to redo the aggregation has a clear practical advantage. This also makes it easier to add or remove samples.

With respect to quality checks, the first issue is the quality of the array measurements, which depends on the success of the experimental steps from RNA preparation to scanning that include sample labelling, hybridization and washing. Quality control (QC) is essential to identify arrays which need to be repeated or might have to be excluded from the analysis. The QC procedures include visual inspection of the image to identify spatial patterns indicative of artefacts as well as quantitative analysis based on sets of quality metrics. In contrast to relative quality metrics such as signal intensities and distribution, absolute measures rely on control probes as, for example, spike-in probes and replicate probes (Kauffmann and Huber, 2010). As an example, the variability across replicate probe measurements facilitates the assessment of intra-array reproducibility. A dedicated R package "arrayQualityMetrics" has been proposed for the quality assessment of microarray data (Kauffmann *et al.*, 2009). For Agilent microarrays, application-specific QC reports are generated when applying the Feature Extraction software. Raw data quality of each array is evaluated using statistical measures largely based on specific control probes designed for the use with the One-Color or Two-Color RNA Spike-in Kits (Agilent Technologies). A set of 10 spike-in transcripts is used to assess the linear dynamic range of the microarray experiment and the reproducibility of replicate probes. Another part of the QC report details outlier statistics and includes display of spatial distribution of the outlier probes for detecting potential regional biases or artefacts (Figure 6.2). In their work on tiling array data, the authors used a simple statistic to capture the quality of the signal for each particular sample into a single summary statistic that consisted of the ratio of the average variance between probes querying the same protein coding region and the overall variance. A value below 0.1, interpreted as 10% of noise, was found to be indicative of high quality, whereas above 0.2, the data was considered of poor quality. Artefacts that generate

168 CHAPTER 6 Array-based approaches to bacterial transcriptome analysis

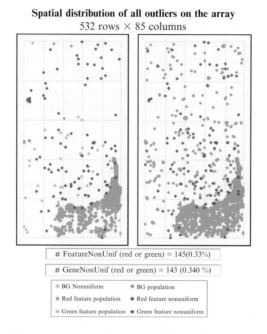

FIGURE 6.2

Graphs from an example QC Report for two-colour gene expression microarrays generated by the Feature Extraction software (Agilent Technologies). The two plots show the spatial distribution of all feature and background (BG) outliers for the green and red channels, respectively. The number (and percentage) of features/genes that are non-uniformity outliers in either the green or red channel are shown below the plots. (For interpretation of the references to colour in this figure legend, the reader is referred to the online version of this chapter.)

misleading expression signals such as the antisense signal, that can arise from the reverse transcriptase activity, need to be assessed specifically.

The second issue relating to quality control concerns the biological consistency of the data. In a typical set of microarray experiments in which a number of biological conditions are tested, several biological replicates are used for each condition. This allows one to verify that their expression signal is more similar within than between conditions. For this purpose, the relationships between samples will typically be summarized into a hierarchical clustering tree (see paragraph on clustering below). If replicates from two or more biological conditions seem to mix consistently, one needs to ask whether this was expected or if it is indicative of a sub-optimal experimental set-up: the applied stress could have been too gentle, for instance. Sometimes, separate groups appear clearly, but some hybridizations seem to be in the wrong group: this may indicate a mislabelling of the samples. Finally, one or a few clear outlier samples may be visible which could indicate an uncontrolled variability. In which case, a closer examination of the genes exhibiting the higher fold changes, for instance, via pairwise log2-scale scatter plots of the data, may identify the origin of the problem and it can be sound to discard a particular sample.

4.2 Expression and differential expression

Even if many, and often a majority, of the genes display a non-ambiguous expression signal, establishing comprehensive lists of expressed versus non-expressed regions from a microarray expression data set turns out to be difficult. The problem stems from background optical and hybridization noises that generate a non-zero baseline imposing a lower limit to detection. In practice, it is not possible to say whether or not a gene is expressed or is expressed below the detection limit. This is inherent to the technology and cannot be compensated for by increasing the number of experiments or probes, but sensitivity differs between platforms. One motivation for pairing each Perfect Match (PM) probe with a so-called Mismatch (MM) probe that differ by a single nucleotide substitution in Affymetrix GeneChip design was precisely to address this issue directly: p-values for a PM signal above the MM signal can be computed and interpreted as a direct assessment of the presence of a transcript. From a number of literature reports, it seems that the benefit of this MM probes was not obvious (see, for instance, Bolstad *et al.*, 2004) and most custom arrays, as well as newer Affymetrix designs, do not include MM probes.

The general procedure is to call "expressed" those genes having an aggregated expression signal that scores above a certain level (Rasmussen *et al.*, 2009), or a distribution of probe-level intensities that differ significantly from an estimated background distribution (Zhou and Abagyan, 2002). Unfortunately, whatever the validity of the approach, the number of called expressed genes will be related not only to the underlying biology but also to the specific signal-to-noise ratios achieved in a particular experiment (or hybridization). Importantly, these technical variations cannot be filtered out by normalization. Fortunately, asking whether a gene is expressed or not may not be the most biologically relevant question when analysing bacterial microarray data sets. Indeed, the current microarray technologies allow for the detection of a consistent expression signal for transcripts with abundance orders of magnitude below one copy per cell. Interpreting this expression signal in terms of coexistence of different subpopulations of cells and/or of background stochastic transcription noise seems therefore more relevant than simply establishing a list of expressed genes.

The situation is different when the purpose is not to compare lists of transcribed/non-transcribed regions between different conditions but to map new transcripts, a question that typically arises in tiling array data analysis. The authors mapped new transcripts using a compendium of 269 tiling array hybridizations by relying on a probabilistic signal smoothing procedure to reconstruct the transcriptional landscape along the chromosome from each hybridization (Figure 6.3; Nicolas *et al.*, 2012). The underlying probabilistic model accounts for abrupt shifts such as transcription breakpoints that correspond to transcription start and termination sites and smaller signal drifts that can reflect artefacts, or reveal expression gradients generated by the interplay of 5′ to 3′ synthesis and "random" termination. For each probe, this led to an expected

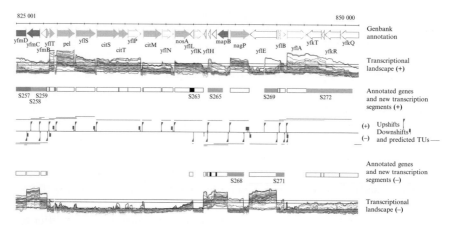

FIGURE 6.3

Bacillus subtilis transcriptional landscape determined by tiling arrays. A sample segment of 25 kbp of the chromosome is shown. The Genbank annotation is displayed above the expression data collected in 269 hybridizations accounting for 104 different biological conditions selected to maximize the diversity of lifestyles (Nicolas et al., 2012). For both strands of the chromosome, 50 transcriptional profiles (coloured curves) illustrative of the diversity observed in the total data set are drawn along with the new transcription segments delimited (coloured boxes). The positions of signal upshifts and downshifts, also represented, were shown to correspond mostly to transcription start sites and terminators and the associated transcriptional units (TUs) were predicted. (For interpretation of the references to colour in this figure legend, the reader is referred to the online version of this chapter.)

transcription signal (in log2-scale) associated with a 95% credibility intervals that integrate out probe affinity effects as measured from genomic DNA hybridization (Nicolas et al., 2009). The lower bound of credibility interval was then compared to a threshold corresponding to $10\times$ the chromosome median, used here as a proxy for the background. Of note, reverse transcription artefacts (limited by addition of actinomycin D) cause well above background expression signals on the antisense strand that cannot be identified by the aforementioned approaches. The authors found that visual inspection of transcriptional landscapes and quantitative measures of the reproducibility of the signal could serve to filter out these artefacts.

Differential expression assessment plays a central role in microarray data analysis. This is particularly true for experimental set-ups comparing a limited number of biological conditions such that the research question can be easily addressed by establishing lists of differentially expressed genes. Hypothesis testing in linear models represents the framework of choice for this purpose with well-established statistical procedures (e.g. analysis of variance, ANOVA) that extend the popular Student *t*-test for comparing the mean of two samples, see, for instance, Smyth (2005). After log-transformation or more general variance stabilizing transformation (Huber et al., 2002), expression levels are written as a sum of terms representing the biological factors, and if necessary technical factors, that explain the relationships between the experiments plus a term of random noise. The p-values (from t-statistics) that quantify the amount of evidence for rejecting a null effect of each factor for a given gene are then computed based on the ratio between fold-change and standard deviation. A p-value (from F-statistic) can also be computed to assess whether any of the factors are associated with a non-zero effect. The linear model framework makes it possible to account for block designs such as time series where measurements at different time-points are made on the same biological replicate.

Despite this pre-existing and well-established framework, the availability of microarray data led to a number of methodological developments to account for the so-called "small n, large p" context where n refers to the number of hybridizations and p to the number of genes. Namely, there are two principal issues that we may want to tackle. The first is that when p-values are computed by treating the genes independently, some information on the variance between biological replicates may be lost if genes tend to exhibit similar behaviour. The loss is greater if the number of replicates used to estimate the variance is small. Several approaches have been proposed to borrow variance information from the whole data set when assessing differential expression for each individual genes. The most popular is the Empirical Bayes method proposed by Smyth (2004) that leads to the so-called moderated p-values. It is based on the assumption that the distribution of variance follows an inverse gamma distribution which may not necessarily be verified (see, for instance, Bourgon et al., 2010). Another approach leading to differential expression statistics that does not rely on such assumption has been proposed and can achieve better gene ranking (Opgen-Rhein and Strimmer, 2007), but assessment of statistical significance becomes non trivial as the method does not directly provide p-values. In practice, any of these methods tend to give more weight to apparent fold-change and less to the apparent standard deviation in the differential expression statistics for each gene, thereby providing an intermediate between fold-change analysis and traditional ANOVA.

The second issue of the "small n, large p" context is the control of the number of false positives arising from multiple testing. If a nominative p-value appropriate for single hypothesis testing such as 0.05 is considered, a large number of genes will be called positive even when none of the genes is differential expressed (around 5% of the genes). Approaches to control this number are referred in the literature as familywise error rate control methods. The best known and simplest being the Bonferroni correction where the p-value cut-off scales inversely with the number of hypotheses tested (genes). The methods that seek to control the number of false discoveries irrespective of the number of true discoveries are, however, typically much more conservative than is usually wanted in exploratory data analysis: in this context, doing 5 false predictions is usually acceptable if this accompanies 95 true predictions as the false discovery rate (FDR) remains limited (5% in this example). A number of methods have been developed to address the control of the FDR by examining the distribution of p-values, and the expected FDR associated with a particular p-value cut-off is often referred to as the q-value. The simplest, conservative, and still very popular method was proposed by Benjamini and Hochberg (1995). More sophisticated and powerful approaches rely on the idea of explicitly estimating the number of true positives and true negatives from the distributions of the *p*-values (Storey, 2002). Finally, some methods have the more ambitious goal of controlling the local FDR which is the expected gene-specific probability of false discovery given the gene *p*-value (see, for instance, Aubert et al., 2004). This may indeed be more relevant from the biologist's standpoint, and an easy-to-use solution is implemented in R library "fdrtool" (Figure 6.4; Strimmer, 2008). These can apply not only to *p*-values but also to any "*z*-score" for which a Gaussian distribution is expected under the null hypothesis.

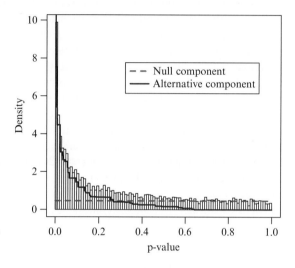

FIGURE 6.4

From the histogram of *p*-values to the estimation of the false discovery rate (FDR). Here, we compared the data collected in two conditions among the 104 conditions investigated in Nicolas et al. (2012): the malate versus glucose carbon sources. Three biological replicates are available for each condition and the *p*-values were obtained using a simple *t*-test. The observed distribution of the *p*-values displays a shoulder near 0 indicative that some genes are differentially expressed because under the null hypothesis of no differences between the two conditions the *p*-values are expected to follow a uniform distribution. The R library "fdrtool" (Strimmer 2008; see main text) is able to decompose the distribution into a mixture of a uniform distribution contributed by those genes that are not differentially expressed (null hypothesis) and a second component resulting from differential expression (alternative hypothesis). From these estimates, the *p*-value cut-off corresponding to a target FDR can be computed. The program also reports the probability of miscalling each gene differentially expressed (the local FDR) given its associated *p*-value. In this example, the alternative component was estimated to account for as much as 54% of the genes (area under the blue curve), but only 11% could be "confidently" called differentially expressed (FDR 5%, green area). (For interpretation of the references to colour in this figure legend, the reader is referred to the online version of this chapter.)

We terminate this section with a few comments on two points related to the interpretation of differential expression analysis. The first point is that with progressively better control of the biological and the technical variance, one can anticipate finding, as soon as the physiological status of the cell differs, a very large number of genes statistically differentially expressed. In this situation, establishing a list of differentially expressed genes would not help to understand the biology and another criterion has to be considered for identifying interesting genes. In practice, one often feels that the list of differentially expressed genes is either uncomfortably small or large. This is not to say that gene-wise differential expression analysis is pointless, interpreting differences in expression levels that are not supported from the statistical viewpoint will often be misleading. When the number of statistically differentially expressed genes is high, careful fold-change analysis and biological considerations would help to disentangle the direct consequences of the difference in biological condition, as opposed to secondary transcriptome modifications. Of note, principled methods to test differential expression above a given fold-change cut-off have been proposed (McCarthy and Smyth, 2009) and may become increasingly relevant as technologies and thus statistical power provided by better experimental set-ups increases. The second point worth mentioning is the concern that the actual level of correlation in expression data might exceed what usual FDR control procedures can accommodate, thereby leading to a wrong feeling of statistical confidence (Qiu *et al.*, 2005). Permutation approaches to FDR control are thus preferred by some authors because of their greater robustness (Tusher *et al.*, 2001; Ge *et al.*, 2003; Grant *et al.*, 2007). It has also been argued that local FDR procedures, working directly with the test statistics rather than with the derived p-values, might be more robust to strong correlation (Strimmer, 2008).

4.3 Towards gene expression networks

Networks provide the most relevant representations to explain many gene expression patterns, with regulatory processes corresponding to network edges and the genes whose expression is monitored correspond to network nodes. A second type of node can be introduced to accommodate quantities that are not directly measured and needs to be inferred, such as transcription factor activities, or other variables that impact on gene expression. However, inferring global regulation networks that map faithfully to the underlying biology is usually not possible from gene expression data alone, due to the impact of missing observations such as post-transcriptional processes, metabolites and environmental variables. In practice, many popular approaches tackle less ambitious problems by introducing more *ad hoc* concepts such as "influential networks" whose edges represent direct and indirect relationships; underlying "expression modes" or "eigengenes" that can be assimilated to transcription factor activities; gene clusters that can be related to actual biological regulons. The second route to network inference from expression data is to focus on a subnetwork and to incorporate prior biological knowledge and additional data sources. In this section, we give a selective overview of the first set of approaches that contribute to the successful mining of transcriptomic data. We refer to thoughtful reviews for the reader interested in more principled network modelling, including boolean, differential equation and Bayesian networks (Hecker *et al.*, 2009; Lee and Tzou, 2009).

The most direct approach to putting expression data in to the context of gene regulatory networks is to analyze gene sets built on prior knowledge of the biological networks. The simplest implementation of this idea is to compare a list of genes established from the analysis of an expression data set to pre-established gene sets representative of regulons, functional categories or metabolic pathways such as the functional classification of *B. subtilis* genes provided by SubtiWiki (http://subtiwiki.uni-goettingen.de/, Mäder *et al.*, 2012). The overlap will be identified as statistically significant on the basis of the Fisher exact test or, equivalently, the hypergeometric model. More sophisticated approaches have been proposed that bypass the need to select a cut-off value defining the list of differentially expressed genes (Nam and Kim, 2008; Ackermann and Strimmer, 2009). These approaches can be useful in situations where the number of genes that could be identified with a reasonable FDR control is too small, as the data collected for a gene set is pooled to provide greater statistical power. The gene set approaches can also be relevant when the number of differentially expressed genes is large (depending on the null hypothesis selected, see classification introduced in Tian *et al.*, 2005) when the focus of the analysis is the comparison of the profiles of differential expression between gene sets rather than the detection of differential expression.

Clustering is the approach by which one seeks to identify groups among elements that, in the context of array experiments, can either be the genes or the experiments (i.e. conditions or individual replicates). As mentioned earlier, the clustering of the experiments is virtually always useful as a data quality check. However, the clustering of the genes is generally more relevant in a systems biology perspective (Grant *et al.*, 2007). The most popular algorithms for clustering rely on the explicit or implicit definition of a distance between elements (most often Euclidean or based on correlation coefficients) and can either optimize the clustering with respect to some criterion for a given number of classes (for instance, *k*-means) or build a hierarchical tree that simultaneously defines clusters for any number of classes (hierarchical clustering). The very popular heatmap representation is obtained by colour coding the expression matrix after reordering rows and/or columns by hierarchical clustering (Figure 6.5, Eisen *et al.*, 1998). Clustering approaches based on information theory (Slonim *et al.*, 2005), mixture models (Ghosh and Chinnaiyan, 2002) and graph theory (Sharan *et al.*, 2003) are also used. Evaluation of the clustering methods from a biology standpoint is difficult and typically relies on measuring the overlap with predetermined gene sets as indicated above (Handl *et al.*, 2005). Probably, the main risk with clustering arises from the availability of a wide diversity of approaches that makes it easy to select the particular procedure that provides a "desired" result instead of a reliable representation of the data. It is indeed crucial to critically assess and understand the results. It is, for example, wise to check the robustness of the conclusions with respect to sampling or to the clustering approach.

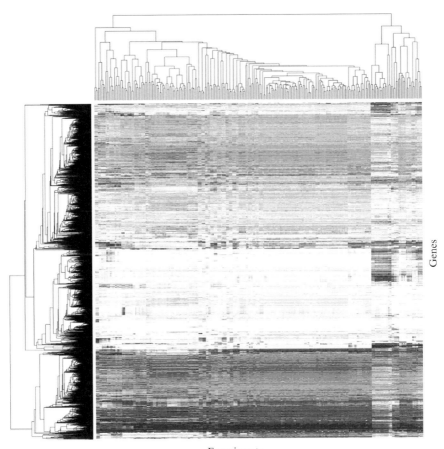

Experiments

FIGURE 6.5

Example of heatmap representation of an expression data set. Here, the heatmap was generated using the data of Nicolas et al. (2012) by reordering both the genes (5875 rows including newly defined transcribed segments) and the experiments (269 columns) according to the results of hierarchical clustering ("average link" on the pairwise Euclidean distance matrix). High expression is represented by darker colours. (For interpretation of the references to colour in this figure legend, the reader is referred to the online version of this chapter.)

The main limitation of the clustering approaches is that they lead to the definition of clusters that do not overlap, whereas the regulons of the different transcription factors or the response to various environmental stimuli do overlap. Several approaches have been proposed to alleviate this constraint. Probably, the most ambitious is the idea of biclustering, whereby attempts are made to find subsets of genes and conditions that show particularly coherent patterns (Madeira and Oliveira, 2004;

Prelić et al., 2006). The rationale behind the bicluster approach is to better match the concept of "regulatory module": a subset of genes that are affected (positively or negatively) in a subset of the conditions. This leads to almost intractable issues relating to both the huge number of possible biclusters and the difficulty of finding relevant criterion to score them.

Another interesting alternative to traditional clustering methods is offered by "component analysis" approaches, in particular, principal component analysis (PCA) and independent component analysis (ICA). PCA and the related singular value decomposition are the simplest and most popular dimensionality reduction techniques that consist of projecting a set of elements (here the genes) from a high-dimensional space (the space of conditions or hybridizations) onto a low-dimensional space (the principal components) (Wall et al., 2003). Principal components, "eigengenes," following the terminology of Alter et al. (2000), are ordered in the sense that the ith principal component captures less variance than the $i-1$th but still as much as possible of the residual variance, that is, what is not already accounted for by all the first $i-1$ principal components. The ICA is a more complicated approach which enforces a statistical independence constraint, instead of orthogonality (uncorrelatedness) in the PCA; it has been argued to be more relevant for biological problems and in practice produces a different decomposition. The underlying factors identified by ICA were termed "expression modes" by Liebermeister (2002). In both PCA and ICA, the coordinates of the new components can serve to define gene clusters that can overlap (Lee and Batzoglou, 2003). Network component analysis is yet another "component analysis" method that cannot be used for cluster discovery as the connectivity matrix which defines whether a particular gene is connected to/regulated by a particular underlying component/transcription factor is given beforehand. Instead, from the activity of the target genes, the method estimates the activity of each transcription factor and the extent of each regulation (Liao et al., 2003).

We end this tour by mentioning the great potential of combining transcriptome analysis and the search for transcription factor binding sites. This combined approach leads closer to gene regulatory network discovery than the previous approaches as a molecular basis for the regulation is proposed alongside a list of regulated genes. Its simplest implementation consists of applying motif discovery algorithms such as MEME (Bailey and Elkan, 1994) or MDscan (Liu et al., 2002), to the promoter regions upstream of genes identified as an expression cluster or as differentially expressed. In this case, a DNA motif may become detectable because the set of sequences subjected to the analysis has been enriched in this motif. Alternative methods, that reverse the perspective of the analysis compared to traditional motif discovery algorithms by trying to model the expression pattern as a function of the sequence instead of focusing on sequence modelling, have also been proposed and show promising results. A clear advantage is that they can more easily bypass the need to delineate the exact contours of each clusters in which a motif is searched for. REDUCE (Bussemaker et al., 2001; Foat et al., 2006), for instance, makes use of a continuous expression trait such as fold-change in a differential expression analysis or coordinates on a principal component axis (Zhang *et al.*, 2008). Another algorithm, FIRE, uses clusters but allows the motifs to extend across several clusters (Elemento *et al.*, 2007). Finally, we mention our own implementation of an idea to combine expression profiles and motif discovery, HMMTREE, whose sequence model accounts for pairwise correlations between promoter activities summarized in a hierarchical clustering tree to discover Sigma factor binding sites (Nicolas *et al.*, 2012).

5 FINAL COMMENTS

The analysis of bacterial transcriptomes with expression microarrays and genomic tiling arrays, respectively, has reached a mature stage with respect to experimental procedures and data analysis strategies. It allows processing of large numbers of samples in a relatively short period of time and at relatively low cost compared to the emerging RNA-seq approach. Gene expression microarrays, although based on an existing genome annotation, are still adequate tools to accommodate systems biology, but with obvious limitations when compared to unbiased whole-transcriptome methods. In particular, RNA-seq technology shows great promise with single nucleotide resolution and higher sensitivity and dynamic range than tiling arrays, even though it is still less well-established and lacks widely accepted standards for data generation and analysis (Mäder *et al.*, 2011). Indeed, the microarray field has led to such significant efforts towards methodological refinements over a period of 15 years that, even if the technology is finally discontinued, understanding the basic principles that have been developed will certainly be of interest for future technologies as it is already true for RNA-seq.

References

Alter, O., Brown, P. O., and Botstein, D. (2000). Singular value decomposition for genome-wide expression data processing and modeling. *Proc. Natl. Acad. Sci. U.S.A.* **97**, 10101–10106.

Aubert, J., Bar-Hen, A., Daudin, J.-J., and Robin, S. (2004). Determination of the differentially expressed genes in microarray experiments using local FDR. *BMC Bioinformatics* **5**, e125.

Ackermann, M. and Strimmer, K. (2009). A general modular framework for gene set enrichment analysis. *BMC Bioinformatics* **10**, e47.

Bailey, T. L. and Elkan, C. (1994). Fitting a mixture model by expectation maximization to discover motifs in biopolymers. *Proc. Int. Conf. Intell. Syst. Mol. Biol.* **2**, 28–36.

Benjamini, Y. and Hochberg, Y. (1995). Controlling the false discovery rate: a practical and powerful approach to multiple testing. *J. Roy. Stat. Soc. Ser. B* **57**, 289–300.

Blom, E. J., Ridder, A. N., Lulko, A. T., Roerdink, J. B., and Kuipers, O. P. (2011). Time-resolved transcriptomics and bioinformatic analyses reveal intrinsic stress responses during batch culture of *Bacillus subtilis*. *PLoS One* **6**, e27160.

Bolstad, B. M., Collin, F., Simpson, K. M., Irizarry, R. A., and Speed, T. P. (2004). Experimental design and low-level analysis of microarray data. *Int. Rev. Neurobiol.* **60**, 25–58.

Bolstad, B. M., Irizarry, R. A., Astrand, M., and Speed, T. P. (2003). A comparison of normalization methods for high density oligonucleotide array data based on bias and variance. *Bioinformatics* **19**, 185–193.

Bourgon, R., Gentleman, R., and Huber, W. (2010). Independent filtering increases detection power for high-throughput experiments. *Proc. Natl. Acad. Sci. U.S.A.* **107**, 9546–9551.

Buck, M. J. and Lieb, J. D. (2004). ChIP-chip: considerations for the design, analysis, and application of genome-wide chromatin immunoprecipitation experiments. *Genomics* **83**, 349–360.

Buescher, J. M., Liebermeister, W., Jules, M., Uhr, M., Muntel, J., Botella, E., Hessling, B., Kleijn, R. J., Le Chat, L., Lecointe, F., Mäder, U., Nicolas, P., Piersma, S., Rügheimer, F., Becher, D., Bessieres, P., Bidnenko, E., Denham, E. L., Dervyn, E., Devine, K. M., Doherty, G., Drulhe, S., Felicori, L., Fogg, M. J., Goelzer, A., Hansen, A., Harwood, C. R., Hecker, M., Hubner, S., Hultschig, C., Jarmer, H., Klipp, E., Leduc, A., Lewis, P., Molina, F., Noirot, P., Peres, S., Pigeonneau, N., Pohl, S., Rasmussen, S., Rinn, B., Schaffer, M., Schnidder, J., Schwikowski, B., van Dijl, J. M., Veiga, P., Walsh, S., Wilkinson, A. J., Stelling, J., Aymerich, S., and Sauer, U. (2012). Global network reorganization during dynamic adaptations of *Bacillus subtilis* metabolism. *Science* **335**, 1099–1103.

Bussemaker, H. J., Li, H., and Siggia, E. D. (2001). Regulatory element detection using correlation with expression. *Nat. Genet.* **27**, 167–171.

Bustin, S. A., Benes, V., Nolan, T., and Pfaffl, M. W. (2005). Quantitative real-time RT-PCR—a perspective. *J. Mol. Endocrinol.* **34**, 597–601.

Canelas, A. B., Harrison, N., Fazio, A., Zhang, J., Pitkänen, J. P., van den Brink, J., Bakker, B. M., Bogner, L., Bouwman, J., Castrillo, J. I., Cankorur, A., Chumnanpuen, P., Daran-Lapujade, P., Dikicioglu, D., van Eunen, K., Ewald, J. C., Heijnen, J. J., Kirdar, B., Mattila, I., Mensonides, F. I., Niebel, A., Penttilä, M., Pronk, J. T., Reuss, M., Salusjärvi, L., Sauer, U., Sherman, D., Siemann-Herzberg, M., Westerhoff, H., de Winde, J., Petranovic, D., Oliver, S. G., Workman, C. T., Zamboni, N., and Nielsen, J. (2010). Integrated multilaboratory systems biology reveals differences in protein metabolism between two reference yeast strains. *Nat. Commun.* **1**, 145.

Chandriani, S. and Ganem, D. (2007). Host transcript accumulation during lytic KSHV infection reveals several classes of host responses. *PLoS One* **2**, e811.

Chomczynski, P. and Sacchi, N. (1987). Single-step method of RNA isolation by acid guanidinium thiocyanate-phenol-chloroform extraction. *Anal. Biochem.* **162**, 156–159.

Churchill, G. A. (2002). Fundamentals of experimental design for cDNA microarrays. *Nat. Genet.* **32**, 490–495.

Dufour, Y. S., Wesenberg, G. E., Tritt, A. J., Glasner, J. D., Perna, N. T., Mitchell, J. C., and Donohue, T. J. (2010). chipD: a web tool to design oligonucleotide probes for high-density tiling arrays. *Nucleic Acids Res.* **38**, W321–W325.

Durand, S., Gilet, L., Bessières, P., Nicolas, P., and Condon, C. (2012). Three essential ribonucleases-RNase Y, J1, and III-control the abundance of a majority of *Bacillus subtilis* mRNAs. *PLoS Genet.* **8**, e1002520.

Eisen, M. B., Spellman, P. T., Brown, P. O., and Botstein, D. (1998). Cluster analysis and display of genome-wide expression patterns. *Proc. Natl. Acad. Sci. U.S.A.* **95**, 14863–14868.

Elemento, O., Slonim, N., and Tavazoie, S. (2007). A universal framework for regulatory element discovery across all genomes and data types. *Mol. Cell* **28**, 337–350.

Eymann, C., Homuth, G., Scharf, C., and Hecker, M. (2002). *Bacillus subtilis* functional genomics: global characterization of the stringent response by proteome and transcriptome analysis. *J. Bacteriol.* **184**, 2500–2520.

Falord, M., Mäder, U., Hiron, A., Débarbouillé, M., and Msadek, T. (2011). Investigation of the *Staphylococcus aureus* GraSR regulon reveals novel links to virulence, stress response and cell wall signal transduction pathways. *PLoS One* **6**, e21323.

Filiatrault, M. J. (2011). Progress in prokaryotic transcriptomics. *Curr. Opin. Microbiol.* **14**, 579–586.

Foat, B. C., Morozov, A. V., and Bussemaker, H. J. (2006). Statistical mechanical modeling of genome-wide transcription factor occupancy data by MatrixREDUCE. *Bioinformatics* **22**, 141–149.

Ge, Y., Dudoit, S., and Speed, T. P. (2003). Resampling-based multiple testing for microarray data analysis. *Test* **12**, 1–77.

Gentleman, R. C., Carey, V. J., Bates, D. M., Bolstad, B., Dettling, M., Dudoit, S., Ellis, B., Gautier, L., Ge, Y., Gentry, J., Hornik, K., Hothorn, T., Huber, W., Iacus, S., Irizarry, R., Leisch, F., Li, C., Maechler, M., Rossini, A. J., Sawitzki, G., Smith, C., Smyth, G., Tierney, L., Yang, J. Y., and Zhang, J. (2004). Bioconductor: open software development for computational biology and bioinformatics. *Genome Biol.* **5**, R80.

Georg, J., Voss, B., Scholz, I., Mitschke, J., Wilde, A., and Hess, W. R. (2009). Evidence for a major role of antisense RNAs in cyanobacterial gene regulation. *Mol. Syst. Biol.* **5**, 305.

Ghosh, D. and Chinnaiyan, A. M. (2002). Mixture modelling of gene expression data from microarray experiments. *Bioinformatics* **18**, 275–286.

Grant, G. R., Manduchi, E., and Stoeckert, C. J. Jr., (2007). Analysis and management of microarray gene expression data. *Curr. Protoc. Mol. Biol.* **77**, 19.6.1–19.6.30.

Hahne, H., Mäder, U., Otto, A., Bonn, F., Steil, L., Bremer, E., Hecker, M., and Becher, D. (2010). A comprehensive proteomics and transcriptomics analysis of *Bacillus subtilis* salt stress adaptation. *J. Bacteriol.* **192**, 870–882.

Handl, J., Knowles, J., and Kell, D. B. (2005). Computational cluster validation in post-genomic data analysis. *Bioinformatics* **21**, 3201–3212.

Hauser, N. C., Vingron, M., Scheideler, M., Krems, B., Hellmuth, K., Entian, K.-D., and Hoheisel, J. D. (1998). Transcriptional profiling on all open reading frames of *Saccharomyces cerevisiae*. *Yeast* **14**, 1209–1221.

Hecker, M., Lambeck, S., Toepfer, S., van Someren, E., and Guthke, R. (2009). Gene regulatory network inference: data integration in dynamic models-a review. *Biosystems* **96**, 86–103.

Herrgård, M. J., Covert, M. W., and Palsson, B.Ø. (2004). Reconstruction of microbial transcriptional regulatory networks. *Curr. Opin. Biotechnol.* **15**, 70–77.

Høvik, H. and Chen, T. (2010). Dynamic probe selection for studying microbial transcriptome with high-density genomic tiling microarrays. *BMC Bioinformatics* **11**, 82.

Hubbell, E., Liu, W. M., and Mei, R. (2002). Robust estimators for expression analysis. *Bioinformatics* **18**, 1585–1592.

Huber, W., von Heydebreck, A., Sültmann, H., Poustka, A., and Vingron, M. (2002). Variance stabilization applied to microarray data calibration and to the quantification of differential expression. *Bioinformatics* **18**, S96–S104.

Huber, W., Toedling, J., and Steinmetz, L. M. (2006). Transcript mapping with high-density oligonucleotide tiling arrays. *Bioinformatics* **22**, 1963–1970.

Irizarry, R. A., Hobbs, B., Collin, F., Beazer-Barclay, Y. D., Antonellis, K. J., Scherf, U., and Speed, T. P. (2003). Exploration, normalization, and summaries of high density oligonucleotide array probe level data. *Biostatistics* **4**, 249–264.

Irizarry, R. A., Wu, Z., and Jaffee, H. A. (2006). Comparison of Affymetrix GeneChip expression measures. *Bioinformatics* **22**, 789–794.

Kauffmann, A., Gentleman, R., and Huber, W. (2009). arrayQualityMetrics—a bioconductor package for quality assessment of microarray data. *Bioinformatics* **25**, 415–416.

Kauffmann, A. and Huber, W. (2010). Microarray data quality control improves the detection of differentially expressed genes. *Genomics* **95**, 138–142.

Khanin, R. and Wit, E. (2005). Design of large time-course microarray experiments with two channels. *Appl. Bioinform.* **4**, 253–261.

Lee, S. I. and Batzoglou, S. (2003). Application of independent component analysis to microarrays. *Genome Biol.* **4**, e76.

Lee, W. P. and Tzou, W. S. (2009). Computational methods for discovering gene networks from expression data. *Brief. Bioinform.* **10**, 408–423.

Leek, J. T., Scharpf, R. B., Bravo, H. C., Simcha, D., Langmead, B., Johnson, W. E., Geman, D., Baggerly, K., and Irizarry, R. A. (2010). Tackling the widespread and critical impact of batch effects in high-throughput data. *Nat. Rev. Genet.* **11**, 733–739.

Lemoine, S., Combes, F., and Le Crom, S. (2009). An evaluation of custom microarray applications: the oligonucleotide design challenge. *Nucleic Acids Res.* **37**, 1726–1739.

Liao, J. C., Boscolo, R., Yang, Y. L., Tran, L. M., Sabatti, C., and Roychowdhury, V. P. (2003). Network component analysis: reconstruction of regulatory signals in biological systems. *Proc. Natl. Acad. Sci. U.S.A.* **100**, 15522–15527.

Liebermeister, W. (2002). Linear modes of gene expression determined by independent component analysis. *Bioinformatics* **18**, 51–60.

Liu, X. S., Brutlag, D. L., and Liu, J. S. (2002). An algorithm for finding protein-DNA binding sites with applications to chromatin-immunoprecipitation microarray experiments. *Nat. Biotechnol.* **20**, 835–839.

McCarthy, D. J. and Smyth, G. K. (2009). Testing significance relative to a fold-change threshold is a TREAT. *Bioinformatics* **25**, 765–771.

Madeira, S. C. and Oliveira, A. L. (2004). Biclustering algorithms for biological data analysis: a survey. *IEEE/ACM Trans. Comput. Biol. Bioinform.* **1**, 24–45.

Mäder, U., Nicolas, P., Richard, H., Bessières, P., and Aymerich, S. (2011). Comprehensive identification and quantification of microbial transcriptomes by genome-wide unbiased methods. *Curr. Opin. Biotechnol.* **22**, 32–41.

Mäder, U., Schmeisky, A. G., Flórez, L. A., and Stülke, J. (2012). SubtiWiki—a comprehensive community resource for the model organism *Bacillus subtilis*. *Nucleic Acids Res.* **40**, D1278–D1287.

Nam, D. and Kim, S. Y. (2008). Gene-set approach for expression pattern analysis. *Brief. Bioinform.* **9**, 189–197.

Nielsen, H. B., Wernersson, R., and Knudsen, S. (2003). Design of oligonucleotides for microarrays and perspectives for design of multi-transcriptome arrays. *Nucleic Acids Res.* **31**, 3491–3496.

Nicolas, P., Leduc, A., Robin, S., Rasmussen, S., Jarmer, H., and Bessières, P. (2009). Transcriptional landscape estimation from tiling array data using a model of signal shift and drift. *Bioinformatics* **25**, 2341–2347.

Nicolas, P., Mäder, U., Dervyn, E., Rochat, T., Leduc, A., Pigeonneau, N., Bidnenko, E., Marchadier, E., Hoebeke, M., Aymerich, S., Becher, D., Bisicchia, P., Botella, E., Delumeau, O., Doherty, G., Denham, E. L., Devine, K. M., Fogg, M., Fromion, V., Goelzer, A., Hansen, A., Härtig, E., Harwood, C. R., Homuth, G., Jarmer, H.,

Jules, M., Klipp, E., Le Chat, L., Lecointe, F., Lewis, P., Liebermeister, W., March, A., Mars, R. A. T., Nannapaneni, P., Noone, D., Pohl, S., Rinn, B., Rügheimer, F., Sappa, P. K., Samson, F., Schaffer, M., Schwikowski, B., Steil, L., Stülke, J., Wiegert, T., Wilkinson, A. J., van Dijl, J. M., Hecker, M., Völker, U., Bessières, P., and Noirot, P. (2012). Condition-dependent transcriptome reveals high-level regulatory architecture in *Bacillus subtilis*. *Science* **335**, 1103–1106.

Opgen-Rhein, R. and Strimmer, K. (2007). Accurate ranking of differentially expressed genes by a distribution-free shrinkage approach. *Stat. Appl. Genet. Mol. Biol.* **6**, e9.

Patterson, T. A., Lobenhofer, E. K., Fulmer-Smentek, S. B., Collins, P. J., Chu, T. M., Bao, W., Fang, H., Kawasaki, E. S., Hager, J., Tikhonova, I. R., Walker, S. J., Zhang, L., Hurban, P., de Longueville, F., Fuscoe, J. C., Tong, W., Shi, L., and Wolfinger, R. D. (2006). Performance comparison of one-color and two-color platforms within the MicroArray Quality Control (MAQC) project. *Nat. Biotechnol.* **24**, 1140–1150.

Perocchi, F., Xu, Z., Clauder-Münster, S., and Steinmetz, L. M. (2007). Antisense artifacts in transcriptome microarray experiments are resolved by actinomycin D. *Nucleic Acids Res.* **35**, e128.

Peters, J. M., Mooney, R. A., Kuan, P. F., Rowland, J. L., Keles, S., and Landick, R. (2009). Rho directs widespread termination of intragenic and stable RNA transcription. *Proc. Natl. Acad. Sci. U.S.A.* **106**, 15406–15411.

Prelić, A., Bleuler, S., Zimmermann, P., Wille, A., Bühlmann, P., Gruissem, W., Hennig, L., Thiele, L., and Zitzler, E. (2006). A systematic comparison and evaluation of biclustering methods for gene expression data. *Bioinformatics* **22**, 1122–1129.

Qiu, X., Brooks, A. I., Klebanov, L., and Yakovlev, N. (2005). The effects of normalization on the correlation structure of microarray data. *BMC Bioinformatics* **6**, e120.

Quackenbush, J. (2001). Computational analysis of microarray data. *Nat. Rev. Genet.* **2**, 418–427.

Rasmussen, S., Nielsen, H. B., and Jarmer, H. (2009). The transcriptionally active regions in the genome of *Bacillus subtilis*. *Mol. Microbiol.* **73**, 1043–1057.

Richter, A., Schwager, C., Hentze, S., Ansorge, W., Hentze, M. W., and Muckenthaler, M. (2002). Comparison of fluorescent tag DNA labeling methods used for expression analysis by DNA microarrays. *Biotechniques* **33**, 620–630.

Ruprecht, R. M., Goodman, N. C., and Spiegelman, S. (1973). Conditions for the selective synthesis of DNA complementary to template RNA. *Biochim. Biophys. Acta* **294**, 192–203.

Schena, M., Shalon, D., Davis, R. W., and Brown, P. O. (1995). Quantitative monitoring of gene expression patterns with a complementary DNA microarray. *Science* **270**, 467–470.

Sharan, R., Maron-Katz, A., and Shamir, R. (2003). CLICK and EXPANDER: a system for clustering and visualizing gene expression data. *Bioinformatics* **19**, 1787–1799.

Slonim, N., Atwal, G. S., Tkacik, G., and Bialek, W. (2005). Information-based clustering. *Proc. Natl. Acad. Sci. U.S.A.* **102**, 18297–18302.

Smyth, G. K. (2004). Linear models and empirical Bayes methods for assessing differential expression in microarray experiments. *Stat. Appl. Genet. Mol. Biol.* **3**, e3.

Smyth, G. K. (2005). Limma: linear models for microarray data. In R. Gentleman, V. Carey, S. Dudoit, R. Irizarry & W. Huber (Eds.), *Bioinformatics and Computational Biology Solutions using R and Bioconductor* (pp. 397–420). New York: Springer.

Sorek, R. and Cossart, P. (2010). Prokaryotic transcriptomics: a new view on regulation, physiology and pathogenicity. *Nat. Rev. Genet.* **11**, 9–16.

Storey, J. D. (2002). A direct approach to false discovery rates. *J. R. Stat. Soc. Ser. B* **64**, 479–498.

Strimmer, K. (2008). A unified approach to false discovery rate estimation. *BMC Bioinformatics* **9**, e303.

Thomassen, G. O., Rowe, A. D., Lagesen, K., Lindvall, J. M., and Rognes, T. (2009). Custom design and analysis of high-density oligonucleotide bacterial tiling microarrays. *PLoS One* **4**, e5943.

Tian, L., Greenberg, S. A., Kong, S. W., Altschuler, J., Kohane, I. S., and Park, P. J. (2005). Discovering statistically significant pathways in expression profiling studies. *Proc. Natl. Acad. Sci. U.S.A.* **102**, 13544–13549.

Tusher, V. G., Tibshirani, R., and Chu, G. (2001). Significance analysis of microarrays applied to the ionizing radiation response. *Proc. Natl. Acad. Sci. U.S.A.* **98**, 5116–5121.

van de Peppel, J., Kemmeren, P., van Bakel, H., Radonjic, M., van Leenen, D., and Holstege, F. C. (2003). Monitoring global messenger RNA changes in externally controlled microarray experiments. *EMBO Rep.* **4**, 387–493.

Velculescu, V. E., Zhang, L., Vogelstein, B., and Kinzler, K. W. (1995). Serial analysis of gene expression. *Science* **270**, 484–487.

Wall, M. E., Rechtsteiner, A., and Rocha, L. M. (2003). Singular value decomposition and principal component analysis. In D. P. Berrar, W. Dubitzky & M. Granzow (Eds.), *A Practical Approach to Microarray Data Analysis* (pp. 91–109). Kluwer: Norwell, MA.

Weng, L., Dai, H., Zhan, Y., He, Y., Stepaniants, S. B., and Bassett, D. E. (2006). Rosetta error model for gene expression analysis. *Bioinformatics* **22**, 1111–1121.

Wernersson, R. and Nielsen, H. B. (2005). OligoWiz 2.0–integrating sequence feature annotation into the design of microarray probes. *Nucleic Acids Res.* **33**, W611–W615.

Winter, T., Winter, J., Polak, M., Kusch, K., Mäder, U., Sietmann, R., Ehlbeck, J., van Hijum, S., Weltmann, K. D., Hecker, M., and Kusch, H. (2011). Characterization of the global impact of low temperature gas plasma on vegetative microorganisms. *Proteomics* **11**, 3518–3530.

Wu, Z., Irizarry, R. A., Gentleman, R., Murillo, F. M., and Spencer, F. (2004). A model based background adjustment for oligonucleotide expression arrays. *J. Am. Stat. Assoc.* **99**, 909–917.

Yang, Y. H., Dudoit, S., Luu, P., Lin, D. M., Peng, V., Ngai, J., and Speed, T. P. (2002). Normalization for cDNA microarray data: a robust composite method addressing single and multiple slide systematic variation. *Nucleic Acids Res.* **30**, e15.

Yu, W. H., Høvik, H., Olsen, I., and Chen, T. (2011). Strand-specific transcriptome profiling with directly labeled RNA on genomic tiling microarrays. *BMC Mol. Biol.* **12**, e3.

Zhang, N. R., Wildermuth, M. C., and Speed, T. P. (2008). Transcription factor binding site prediction with multivariate gene expression data. *Ann. Appl. Stat.* **2**, 332–365.

Zhang, W., Li, F., and Nie, L. (2010). Integrating multiple 'omics' analysis for microbial biology: application and methodologies. *Microbiology* **156**, 287–301.

Zhou, Y. and Abagyan, R. (2002). Match-only integral distribution (MOID) algorithm for high-density oligonucleotide array analysis. *BMC Bioinformatics* **3**, e3.

Index

Note: Page numbers followed by "*f*" indicate figures, and "*t*" indicate tables.

A

Acinetobacter spp., 116
Aequorea victoria, 1–2, 3–4
Amino acid biosynthetic pathway, 8
Array-based approach
   ad hoc concept, 173–174
   Affymetrix GeneChip array, 165–166
   aggregation method, 167
   Agilent Technologies and Roche NimbleGen platform, 155
   array fabrication method, 153
   array hybridization and data extraction, 164–165
   arrayQuality-Metrics, 167–168
   Bacillus subtilis transcriptional landscape, 169–170, 170*f*
   bacterial metabolism and regulation, 151
   bicluster approach, 175–176
   Bonferroni correction, 171–172
   cDNA synthesis and labelling, 160–164
   chipD, 155
   clustering, heatmap, 174, 175*f*
   CRAN, 155–156
   empirical Bayes method, 171
   experimental design, 156–158
   FDR estimation, 171–172, 172*f*
   Fisher exact test, 174
   fold-change analysis, 173
   gel-like image and electropherograms, RNA, 158, 159*f*
   GeneChip array, 154
   genome-wide expression analysis, 151–152
   genomic tiling array, 152
   hierarchical clustering tree, 168–169
   high-quality RNA preparation, 158
   HMMTREE, 176–177
   invariant set normalizations, 166
   linear model framework, 170–171
   OligoWiz 2.0, 154–155
   PCA and ICA, 176
   PM and MM probe, 169
   probe, thermodynamic property, 154–155
   protein-DNA interaction, 152–153
   REDUCE and FIRE, 176–177
   RMA approach, 165
   RNA Clean-Up and Concentration Kit, 158–160
   RNA-seq approach, 177
   in situ synthesized array, 153–154
   spike-in normalization, 166
   spotted array, limitation, 154
   strand-specific cDNA generation and labelling, 160*b*
   transcriptome study, 152
   two-colour gene expression microarray QC, 167–168, 168*f*
   whole-transcriptome study, 152
Artificial neural network (ANN)
   advantages and disadvantages, 49
   application, 49–50
   multi-layer perceptron, 48, 48*f*
   sigmoid function, 48, 49*f*

B

Bacillus subtilis
   amyE, 108
   dual-labelling, 108–109
   plasmid maps, 108, 109*f*, 110*t*
   pSG-and pNG-based vector, cloning, 109–112
   pYG1, 112
   screening transformants, 113
   single-and double-crossover integration, 108, 111*f*
   transformation, 112–113

C

Cluster analysis
   application, 45
   cluster, number, 46
   dendrogram, hierarchical clustering algorithm, 45, 46*f*
   Euclidean distance, 43–44
   guilt-by-association approach, 44–45
   Hamming distance, 43
   k-means, 44
   k-nearest neighbour (k-nn) algorithm, 44
   microarray analysis pipeline, 45–46, 47*f*
   ordering, sensitivity, 46
   stochasticity, 46

Cluster analysis (*Continued*)
  visual representation, 35*f*, 47
  yeast, two-hybrid dataset, 42–43
Corynebacterium glutamicum, 131

D

Data mining
  algorithm, conceptual hierarchy, 29, 30*f*
  business and marketing, 28
  CI-based algorithm, 68
  database rot, 63
  data integration and network analysis
    BioGrid, 57–58
    edge confidence estimation, 57
    interactome analysis, 58
    motifs, 58–59
    SinI/R operon, *Bacillus subtilis*, 56, 57*f*
    transcriptional network regulation, 59
  data manipulation, 31–32
  data source, 30
  definition of, 28
  discriminant analysis, 33–35, 35*f*
  DNA sequence and protein complex formation, 27
  domain knowledge, 30
  eScience
    cloud computing, 60–61
    crowd sourcing, 62
    grid computing, 59–60
    haemolytic uraemic syndrome, 62
    Taverna workflow, 60, 61*f*
    web services, 60
  fishing expedition, 29
  gene ontology and stepwise linear discriminant analysis, 29
  inductive logic programming, 63–64
  interactomes, 64
  life cycle, 30, 31*f*
  linear regression, 33, 34*f*
  machine learning technique
    ANN (*see* Artificial neural network (ANN))
    cluster analysis (*see* Cluster analysis)
    decision tree algorithm (*see* Decision tree algorithm)
    HMM, 38–39, 39*f*
    ontologies and text mining (*see* Ontology and text mining)
    statistical approach, 37
    SVM, 37–38
  metagenomics data, 65
  microbial behaviour, 68
  noise, 65–66
  overfitting, 66–67, 66*f*
  PCA, 35–37, 36*f*
  peaking phenomenon, 67, 67*f*
  pre-processing step, 31
  Saccharomyces cerevisiae, 27–28
  statistical analysis, 32, 33*t*
Decision tree algorithm
  advantages of, 42
  application, 42
  C4.5 algorithm, 40, 42
  entropy, 40–41, 41*f*
  MALDI-TOF MS, *Salmonella enterica*, 40, 40*f*
Dictyostelium discoides, 51–52

E

Empirical Bayes method, 171
Escherichia coli, 6–7
  l Red system, 114–116, 115*f*
  pCP20, 116
  PCR construct transformation, 114
  plasmid cloning vector, 113
  primer, gene of interest, 114

F

Fluorescent protein fusion
  Acinetobacter spp., 116
  agarose slide preparation, 119
  Bacillus subtilis
    amyE, 108
    dual-labelling, 108–109
    plasmid maps, 108, 109*f*, 110*t*
    pSG-and pNG-based vector, cloning, 109–112
    pYG1, 112
    screening transformants, 113
    single-and double-crossover integration, 108, 111*f*
    transformation, 112–113
  Escherichia coli
    l Red system, 114–116, 115*f*
    pCP20, 116
    PCR construct transformation, 114
    plasmid cloning vector, 113
    primer, gene of interest, 114
  functional analysis, 117
  GFP, 107–108
  growth condition
    liquid, 118
    solid, 118
    subcellular domain, 117–118
  image processing, 122–124, 122*f*

imaging hardware
  filter properties, 120–121, 121t
  microscope and camera, 119–120
  objective lens, 120
Imaris and ImageJ, 121
MetaMorphO, 121
protein dynamics, 124
protein localisation and abundance, 107
signal comparison, 123–124, 123f
Staphylococcus aureus, 116
Tet array and pLau53, 117

G

β-Galactosidase, 3
Green fluorescent protein (GFP), 1–2, 3–4

H

Hidden Markov model (HMM), 38–39, 39f
High-resolution temporal analysis
  amino acid biosynthetic pathway, 8
  cloning-specific promoter, 7
  data analysis, 20–21
  dynamic gene expression model, 22
  E. coli and *S. typhimurium*, promoter activity, 6–7
  E. coli, transcriptional promoter fusion, 9
  flagellar synthesis, 7
  fragment selection and generation, 12–13
  gene fusion technology, 2–3
  GFP fusion technology, 9–11
  growth and data collection, 19–20
  heat map, visualization, 21–22
  iron-responsive promoter, *Salmonella enterica*, 7
  ligation-independent cloning, 13, 14f
  Lux and GFP, 1–2
  naphthenic acid, 9
  pAH321, 15
  pBaSysBioII, 13
  pFSB79, 15
  pGFPamy, pCFPamy, pYFPamy/pGFPbglS, pCFPbglS, pYFPbglS, 13–15
  and plasmid preparation, 15–16
  promoter activity, 6
  reductionist approach, 1
  regulatory network, reorganization, 11
  reporter protein
    β-galactosidase, 3
    GFP, 3–4
    luciferase, 4–5
  RNA-Seq and microarray, 1–2
  SOS DNA repair system, 7–8
  spo0A transcription fluctuation, 11–12
  strain storage
    expression profile, 18–19
    multiplexed manipulation, 16–18, 17f
    96-well plate, test strain distribution, 16–18, 19f
  transcriptional fusion, 5–6

L

Luciferase, 4–5

M

Mass spectrometry
  ESI, 137
  ESI-MS/MS, 139–140
  ion-pairing chromatography, 138–139
  ion-pairing LC triple quadrupole MS method, 137
  LC-MS/MS data, peak integration, 140–141
Metabolomics
  bacterial cultivation
    inoculum, 130
    M9 minimal medium, 129
    shake flask cultivation, 130
    standard cultivation protocol, 128–129
  biomass, 145–146
  cold methanol quenching, 131
  extraction method
    acidic acetonitrile extraction, 136–137
    cold methanol extraction, 135
    hot ethanol extraction, 135–136
  fast filtration protocol, 131–133, 132f
  genome-scale metabolic model, 127
  high-throughput sampling, 134–135
  isotope ratio-based quantification
    bioreactor, 141
    E. coli, anaerobic batch cultivation, 143
    extraction, 143
    fermentation, 142
    inoculum, 142
    metabolite standard mixture, 144
    sampling and quenching, 142
    spiking, 144
    U-^{13}C-glucose minimal medium, 141
    U-^{13}C-labelled glucose, 141–145
  mass spectrometry
    ESI, 137
    ESI-MS/MS, 139–140
    ion-pairing chromatography, 138–139
    ion-pairing LC triple quadrupole MS method, 137
    LC-MS/MS data, peak integration, 140–141
  omics method, 128
  precaution, filtration, 132

Metabolomics (*Continued*)
   pros and cons, 128
   quantification level, 146, 146*t*
   targeted and untargeted metabolic profiling, 127
   thermodynamic analysis, 147
   ultrahigh-throughput analysis, 147
   whole cell broth extraction, 133–134
Munich Information on Protein Sequences (MIPS) database, 27–28

O

Ontology and text mining
   application, 55–56
   BacillOndex, 51, 52*f*
   cloud computing, 53
   computational modelling and theory, 52–53
   definition of, 50–51
   MeSH, 53, 54*f*
   NLP, 53
   OBO, 51–52
   semantics, 55
   structured vocabulary, 51
   TextPresso, 55

P

Principal components analysis (PCA), 35–37, 36*f*
Proteomics
   AQUA method, 85
   best flyer peptides, 86
   cell count determination, 89
   cell dimension estimation, 89–90
   cell disruption, 90, 91*f*
   cultivation, 89
   data-dependent mass spectrometry-based proteomic, 84
   data-independent acquisition, 83–84
   2D gel electrophoresis, 81–82
   differential 2D gel image analysis
      advantages of, 97–98
      fluorescent dye, 98, 98*f*
      MALDI MS, 98–99
   fusion protein-based global scale protein quantification, 84–85
   gel-free proteomics, 82–83
   heavy isotope-labelled internal standard, 85–86
   isotope dilution, 85
   large-scale absolute quantification, 87
   MS^E, 84
   omics technique, 81
   protein concentration determination, 91–92, 92*f*
   QconCAT technique, 86
   sample preparation, 87–89, 88*f*
   SRM-calibrated 2D PAGE, 99–101
   targeted mass spectrometry
      AQUA method, 94–96
      de novo gene design, QconCAT, 95–96
      digestion efficiency evaluation, 93–94, 94*f*
      endogenous peptides *vs.* heavy-labelled spiked peptides, 94
      precursor/fragment ion pair, 93
      QconCAT protein expression and purification, 96
      quantitative SRM assay, 96–97
      quantotypic peptide, 93
      SRM, stable isotope dilution concept, 92–93
   targeted proteomics, 83
   triple quadrupole mass spectrometer, 83

R

Robust Multichip Average (RMA) approach, 165

S

Saccharomyces cerevisiae, 27–28
Salmonella enterica, 7
Salmonella typhimurium, 6–7
Staphylococcus aureus, 116
Support vector machine (SVM), 37–38

T

Targeted mass spectrometry
   AQUA method, 94–96
   de novo gene design, QconCAT, 95–96
   digestion efficiency evaluation, 93–94, 94*f*
   endogenous peptides *vs.* heavy-labelled spiked peptides, 94
   precursor/fragment ion pair, 93
   QconCAT protein expression and purification, 96
   quantitative SRM assay, 96–97
   quantotypic peptide, 93
   SRM, stable isotope dilution concept, 92–93